COMPUTERS AND CHAOS

Conrad Bessant

SIGMA PRESS – Wilmslow, United Kingdom

Copyright ©, Conrad Bessant, 1992

All Rights Reserved. No part of this publication may be reproduced, stored in a retrieval system, or transmitted in any form or by any means, electronic, mechanical, photocopying, recording or otherwise, without prior written permission.

First published in 1992 by

Sigma Press, 1 South Oak Lane, Wilmslow, Cheshire SK9 6AR, England.

British Library Cataloguing in Publication Data

A CIP catalogue record for this book is available from the British Library.

ISBN: 1-85058-247-5

Typesetting and design by

Sigma Hi-Tech Services Ltd

Distributed by

John Wiley & Sons Ltd., Baffins Lane, Chichester, West Sussex, England.

Acknowledgement of copyright names

Within this book, various proprietary trade names and names protected by copyright are mentioned for descriptive purposes. Full acknowledgment is hereby made of all such protection.

Printed and bound in Great Britain by
Dotesios Ltd, Trowbridge, Wiltshire.

Preface

With the dramatic increases in home computer performance over the past few years it is not surprising that chaos theory and fractal graphics programming have become extremely popular. People can now create, in their own homes, the programming experiments and images which only 10 years ago were leading edge scientific research. The Atari ST is an ideal system for exploring this exciting field, as it can perform the necessary calculations very quickly and the output can be displayed using high resolution graphics.

This book details the methods used to produce most important fractals and also provides a wide variety of ideas for further experimentation, using many of the ST's most powerful features including colour graphics and even sound. The concepts behind these images are discussed using only the most elementary maths and the minimum of technical jargon. Wider aspects of chaos, such as its history and notable applications in the real world, have also been included.

Fully annotated GFA BASIC example programs are given throughout to illustrate the concepts and to provide a base for experimentation. Such examples can even be appreciated by novice programmers who will find this an enjoyable way to learn more about ST programming techniques. Appendices provide useful GFA BASIC and C routines and give hints for converting the programs to other ST languages.

The layout of this book is such that it may be used either as a practical introduction to chaos or as a reference text for the experienced reader, where chapters may be skipped or read out of sequence if necessary.

I must thank my parents, Ange and Mike, for their support and assistance during the writing of this book. Also thanks to Gemma for distribution of the support disk, and to Russell for his help with the initial experiments.

Compatability note: The examples given were written using GFA BASIC 2.0, which is compatible with later versions. All of the programs will work on any ST with any monitor configuration. However, if you have an Atari STE or TT you will need to use GFA BASIC 3.5, as it would appear that version 2.0 is incompatible with these two machines. The GFA BASIC compiler is not essential to this book, but it can be used to improve the performance of all the example programs.

Conrad Bessant

SUPPORT DISK

A support disk for this book is available direct from:

Appletrees
8, The Marsh
Carlton
Bedfordshire MK43 7JU

The price is £8.99.
Cheques and postal orders should be made payable to *Gemma Bessant*.

The disk contains all of the listings from the book in a form suitable for loading into any version of GFA BASIC higher than 2.0. Also included are three very fast pre-compiled fractal exploration programs, each with a full GEM interface, file handling facilities and comprehensive documentation. These three programs can be used to explore Mandelbrot/Julia sets, the Feigenbaum diagram, the May equation and fractal plants, curves and flakes.

CONTENTS

1. Introduction 1
   Chaos 2
   Fractals 3
   The Atari ST 4
   How to Use this Book 5
      Monitor Compatibility 5
      Format of Equations 5
      Program Listings and the Function Library 6
         The Function Library 7
         Line Length 7
         Remarks 7
      Screen Dumps 7
      Entering and Debugging the Examples 8

2. Order and Chaos are Related 9
   The Sierpiñski Triangle 9
      A Fractal 13
   Population Dynamics and the Feigenbaum Diagram 15
      Applying the Equation 18
      Enhancements to the Feigenbaum Program 26
   Conclusion 28

3. Weather, Chemistry and Strange Attractors 29
   The Butterfly Effect 35
   The Lorenz Attractor 40
   The Rössler Attractor 42
   Enhancements to the Attractor Programs 43

4. The Mandelbrot Set 45
   The Circle Method 46
      Adding Colour 56
   The Complex Number Method 59
      Complex Numbers 59
      Applying Complex Numbers to the Equation 61

Manipulating the Mandelbrot Set.................................63
   Zooming In 63
   Other Ways of Displaying the Mandelbrot Set 68
Enhancements to the Mandelbrot Programs........................71
   Internal Structure 71

5. Julia Sets...72
The Julia Process..72
   Adding Contours..74
   Derivation from the Complex Number method....................75
A Julia Set Explorer Complete with Bells and Whistles 75
   Planning 76
   Universal Programming Techniques 76
   The Menu Bar 77
   Menu Handling...78
      Desk_About...79
      Dummy 80
   File Operations 80
      File_load 83
      File_save 83
      File_delete 84
      File_Quit 84
      Picture_julia 84
      Picture_zoom 84
      Picture_parameter 85
      Colour...85
   Procedures Not Called Directly from the Menu Handler 86
      Plot_set 86
      Check_res 87
      Round 87
      Open_window and Close_window 87
      Memory_initialise, palette_store and memory_free 88
Enhancements to the Julia Program 95
   Altering the Program for Mandelbrot Exploration.................96
   A Resume Drawing Option 96
   An Intelligent Search Path 96

6. Imitating Nature – Plants, Shrubs and Trees 98
What is a Plant? 98
Describing a Plant 100
Turning Descriptions into Drawings................................102
String Generation...106
Trees 109
When is a Plant Not a Plant?......................................113
   The C-Curve 114
   The Koch Curve..116
Further Experimentation..121
   Branches with Variable Thickness 121
   More Complex Complexity 121
   Using C..122

Contents

7. Imitating Nature – Fractal Landscapes 123
   Isometric Drawing 123
      Using the Landscape Procedure 129
      Improving the Landscape Procedure 130
   Generating Pseudo-Natural Landscapes 131
      Inheritance 132
      Faulting 135
   Using Landscape Techniques to Plot Fractals in 3D ... 138
      The Feigenbaum Diagram 139
      The Mandelbrot Set 140
   Further Ideas for Fractal Landscapes 142
      Shading 143
      Trees and Bushes 143

8. Imitating Nature – Cell Culture 145
   The Martin Process 145
   Return of the Butterfly 151
   Further Experimentation with the Martin Program 151
   Extensions to the Martin Program 152
      Colour 152
      Sound 153
      Compilation 154
   Further Experiments with Natural Fractals 154

9. The Future 156
   Can the Future of Chaos be Predicted? 156
      Fractal Maths in Data Compression Applications .. 156
      Telephones with Minds of their Own 157
      The New Art 158
   What Use is Chaos? 159
      Derivation of Pi Using the Monte Carlo Method .. 159
   Life After Computers and Chaos 163
      Two Headed Mandelbrots and Similar Feigenbaums . 163

Appendix A: Useful Routines 169
   Using the GFA BASIC Procedure Library 169
      Procedure Definitions for Example Programs 170
         Check_res 170
         Fplot 170
         Fdraw 171
         Waitmouse 171
      Screen Dumps 171
      Loading and Saving Pictures 174
      Manipulating Degas Format Picture Files 174
   Useful C routines 176

Appendix B: Mathematics in GFA BASIC 179
   Simple Algebraic Principles 179
      Operator Priorities 180

    Ratios 180
        Fractions as Percentages 181
   Graphs and Co-ordinate Geometry................................... 182
   Indices 184
        Square Roots 184
   The MOD Operator 185
   Angles – Degrees and Radians 187
   Peculiarities of Right-angled Triangles............................ 187
        Pythagoras' Theorem....................................... 188
        Simple Trigonometry....................................... 188

Appendix C: Using Other ST Languages 190
   Hisoft BASIC 191
   STOS BASIC 191
   ST BASIC 191
   The C Language 191

Appendix D: Use of ST Peripherals 193
   Monitors.. 193
   Printers.. 194
        Nine Pin 196
        Twenty-four Pin 196
        Colour... 199
        Plotters 199
   Extra RAM... 199

Bibliography... 200

1

Introduction

Until a few centuries ago it was believed that nature was a totally unpredictable force, and that forecasting or altering natural events was impossible. Religions were born to explain away this unpredictability, and indeed the thought of a great being who arbitrarily decided what was going to happen in the universe fits in well, as it is a known fact that human actions are often unforeseeable.

During the 16th and 17th centuries scientists such as Galileo, Brahe and Kepler combined observational and academic skills in order to determine that the planets of the solar system had cyclic motions, and also devised methods of precisely calculating the future positions of planets. Newton later determined more general laws governing motion, allowing the results of forces and collisions on smaller bodies to be predicted. Naturally the theories of these scientists came into conflict with the religious movements, that still believed the world was flat, and that it lay at the centre of the solar system.

The Church put a lot of effort into silencing these scientists, but gradually their beliefs became widely accepted and society now appreciates that the results of many natural events can be predicted. By the 18th century scientists were becoming complacent, believing that there were very few things on Earth that they could not understand. However, at the beginning of the 20th century scientists were questioning the validity of some of the old theories and equations, and indeed many modifications were made. Even so, scientists were still not content with the accuracy of some of these predictions.

During World War Two the importance of accurate weather forecasts became acutely apparent, and in the following decades scientists were encouraged to become more involved in meteorology, the science of weather and climate. It was popularly hoped that with the new technology available, and with large research funds, accurate long-term weather forecasting would be possible. Despite well financed and well co-ordinated schemes in Europe and the United States this hope was never realised.

At the time people were surprised at the scientists' apparent failure, and could not understand why mundane weather forecasting was so elusive when something as distant and obscure as planetary motion had been predicted 300 years previously. A simple hypothesis for the lack of success is that the weather is of a much more complicated nature than those systems for which rules have been found, and it is influenced by a wide variety of outside events. This is a fair assumption for weather, but unpredictable processes do not have to be complicated. For example the flow of water in a pipe, the path of energetic smoke particles in a container and even the action of a simple pendulum bob are hard to forecast in some circumstances.

An answer closer to the truth is that unlike predictable processes, such as the ebb and flow of the tide, the day-to-day weather is not cyclic, it does not repeat to a constant rhythm. Cyclic processes are easier to forecast mainly because a simple model can be built around data collected from previous cycles. However, examples contradict this again, as useful rules have been found for forecasting the results of all manner of events, from collisions to chemical reactions. Naturally, if it was possible to accurately predict the effects of things like the weather, the paths of oil slicks and fluctuations of the world money markets many problems could be solved, and much effort saved. Currently, however, general trends are all that can be found in these important areas.

Scientists working on meteorology theory, such as Edward Lorenz, began research into why weather prediction was so difficult, using simple computer-based mathematical models. Other researchers looked at different unpredictable elements of nature. For example Robert May concentrated on population dynamics, and Benoit Mandelbrot discovered unpredictability in pure mathematics. As people like these published papers of their work connections began to appear between the different kinds of irregularity being investigated. Common characteristics could be seen to be exhibited in erratic processes taking place in all manner of different subject areas, from economics to chemistry. Processes displaying these certain characteristics became known as chaotic processes due to their pseudo-random nature, and the study of their similarities became the new science: chaos.

Chaos

For hundreds of years scientists had disregarded the intricacies of the non-cyclic, erratic side of nature, believing it to be too difficult to predict and managing instead with generalised behaviour patterns. With the birth of chaos theory it seemed as if a solution to these fundamental problems may at last be in sight. Surely if the cause of unpredictability could be found it would be possible to cure it. Some chaos crusaders even rated it as the third great physics discovery of the 20th century, after quantum mechanics and the theory of relativity.

The backbone of chaos theory is the notion that complex behaviour does not have to come from complex rules, thus inferring that a system as complicated as the weather may be summed up in one simple equation. The hope is that chaos will be able to give us an equation for everything. This does, however, seem far from likely at the present time as the science is still in its infancy. What a lot of people fail to understand is why it has taken so long for scientists to acknowledge the importance of chaos. Surely if it is so fundamental it could have been appreciated centuries ago.

The catalyst for the research has undoubtedly been the computer. The millions of calculations that must be performed in even the simplest of chaotic processes make manual calculation practically impossible. In the beginning it was only well-funded defence establishments and universities that could afford the equipment necessary to carry out such experiments. But since the early 80s the price and size of computers have fallen dramatically, while power and availability have increased, bringing chaos within easy reach of many other people. Never before have so many people been able to become actively involved in such a state-of-the-art science, and this has been a major factor in its increasing popularity.

Many books have been written on the subject at all levels and mathematical periodicals, not normally touched with a barge-pole if possible, strangely went missing from reference libraries across the world. This incredible level of interest has grown ever since, with some chaos books becoming instant best-sellers, and less than five years after the discovery of the Mandelbrot set, popular computing magazines were publishing polished programs to display it on home computers.

Fractals

Fractals are abstract shapes existing in two or more dimensions, whose incredible complexity prevents them from being treated like normal geometrical shapes. Fractals have surprisingly little ordered structure, despite the fact that they originate from very simple rules. Such images have been irreversibly linked to chaos theory because they are such a good example of how immensely complicated output can be created from a very simple process.

The typical example of a computer generated fractal is the Mandelbrot set, whose infinitely complex structure is uncharacteristic of the simple equation from which it originates. Like most things in chaos theory these computer images are not abstract from the real world, frost on a window and the structure of a tree are natural fractals. Fractals don't always have to be static though. For example, a November 5th sparkler produces a constantly changing fractal structure of sparks.

This is a very generalised definition of a fractal, and a more complete one will be built up as the book progresses. Fractals are most successfully described with the aid of practical experiments and examples, such as those provided for the ST in later chapters.

The Atari ST

In terms of power, price and software availability the Atari ST is undoubtedly one of the best value computers around. It is particularly suitable for fractal work because it can perform the necessary calculations very quickly and the output can be displayed in up to 16 colours, or up to a resolution of 640x400 pixels. The inclusion of the mouse, printer port and disk drive mean that chaos programs can be made very user-friendly, with easy storage for the final output.

The only problem with the ST is that, unlike many other popular machines, it does not have a standard language which fully exploits its power. The decision regarding the best language on which to base this book was therefore a difficult one. Machine code was ruled out immediately because of its unstructured, often difficult to understand nature. The C language is widely used by chaos researchers due to its speed, mathematical capabilities and structure. However, versions of C for the ST generally cost about five times more than comparable BASICs, meaning that few ST owners have had a chance to become familiar with it. The only other well supported high-level language for the ST is BASIC, so the choice was narrowed down to the five main ST versions of the language, shown in Table 1.1.

Table 1.1: Popular dialects of BASIC for the Atari ST

Language	Vendor	1990 Price For Interpreter
ST BASIC	Atari	Free with the Atari ST
STOS	Mandarin	19.95*(free with some ST packs)
GFA BASIC 2.0	GFA Data Media	19.95(price includes compiler)
GFA BASIC 3.5	GFA Data Media	49.95
Power BASIC	Hisoft	49.95(price includes compiler)

* *includes various utilities such as music and sprite editors*

Due to its pricing and popularity GFA BASIC 2.0 appeared to be the most suitable choice, and at the time of writing many popular ST magazines were supplying free copies of the interpreter from the package (without manual or compiler). The more expensive BASICs were ruled out, and STOS was not considered as it is geared more towards games than mathematical applications. ST BASIC would have been a natural choice, as everyone has a copy of it, but unfortunately it is far too slow for fractal work. Appendix C gives advice on converting the GFA BASIC example programs to other dialogues of BASIC and to the C language.

I am assured that GFA BASIC 2.0 is fully compatible with all subsequent releases, so if you have version 3.0 or above all the examples should work without alteration. If you have the compiler it is recommended that you use it once you are sure that the program has been typed in correctly (i.e. when it works with the interpreter). On average the programs in this book will run about three times faster after compilation, and speed may be further increased by making minor changes to the given code.

How to Use this Book

Chaos theory can appear daunting, but the use of simple programming procedures and notational conventions in this book should make it very accessible. The following few pages provide details of the conventions used and give other hints to help you get the most out of both the text and the listings.

Monitor Compatibility

The design of the ST means that it is difficult for any one program to fully utilise the features of high resolution monochrome and low/medium resolution monitors or televisions. GFA BASIC supports both monitors, and in order to make this book relevant to as many users as possible all of the examples contained in it are compatible with both configurations.

Every example program in this book calls a procedure called check_res before doing anything else. This determines the screen mode and sets a variable, called monitor, accordingly. This variable is used for scaling purposes when programs are plotting points and drawing lines. If you are using a high-resolution monochrome monitor, such as the Atari SM124, the monitor variable will be set to 2. If you are using a colour monitor or a television you should enter GFA BASIC in low-resolution mode, and check_res will then set the monitor variable to 1.

To save further confusion medium resolution is not supported - check_res will generate an alert box conveying the fact if the program is executed in this mode. The monitor variable does not affect the colour of the output, as most of the programs produce monochrome images, but it does alter the level of detail shown in the plots and the number of calculations performed (higher resolution plots need more detailed calculation).

Format of Equations

Many of the fractals in this book are based on mathematical equations. Such equations are shown using programming syntax wherever possible, instead of normal mathematical notation, to provide a solid link to the example programs and allow non-mathematicians to get to grips with the concepts more readily. A summary of the most commonly used conventions is shown in Table 1.2. If you are an experienced programmer you will probably already be familiar with most of these, but the table should still prove useful when cross-referencing to other chaos literature.

Table 1.2: Conventional mathematical symbols and their BASIC counterparts

Convention (in this book)	Description	Convention (normal maths)
+	add	+
-	minus/negative	-
*	multiply	x
/	divide	÷
1/2	a fraction less than 1	1/2
2³/₄	a fraction larger than 1	2³/₄
=	equal to	=
≡	equivalent to	≡
∝	proportional to	∝
<=	less than or equal to	≤
>=	greater than or equal to	≥
<>	not equal to	≠
2^2	a number raised to a power	2^2
2SQR(16)	square root of a number	$\sqrt{16}$
(1+2)	bracketed expression	(1+2)
((1+1)*2)	multi-bracketed expression	[(1+1)x2]

These operators can generally be said to apply to all the major ST programming languages. However, the function used to raise a number to a power varies considerably, for example in C the `pow()` function is usually used. It is important to note that the 'equivalent to' and 'proportional to' symbols are not implemented in most languages as their definitions are rather too flexible.

Although variable names will be kept consistent between text and program listings their appearance may vary. This is because in mathematics and most computer languages variables are conventionally shown in lower case, whereas GFA BASIC automatically converts the first letter of every variable name to upper case as it is entered into the editor. Also, some variables will be shown in the text with subscripts in order to differentiate between variables with similar characteristics, for example p_{new} and p_{old}. Because BASIC doesn't allow subscripts the variables will be shown in the listings as `pnew` and `pold`.

Program Listings and the Function Library

Great effort has gone into keeping the program listings as short as possible. This has been done to increase clarity, to reduce the possibility of typing errors, and to allow output similar to that shown in the figures to be created with as little effort as possible.

The Function Library

Some of the reduction in length has been brought about by removing commonly used functions from the listings, and placing them in a 'function library' in Appendix A. Routines such as the resolution sensitive `fplot` and `fdraw` routines are included here. Because vital procedures are missing from the listings they will generate a *Procedure not found error* if typed in *verbatim*, and GFA BASIC will return to the editing screen with the cursor at the offending line. Such a line may read as follows:

`@Waitmouse`

In this case the `waitmouse` function will also need to be included in the program, either by typing it in or by merging it from a pre-written file containing all the common functions. Comprehensive advice on the best way to include the relevant functions is given in Appendix A. This may seem rather complicated at first, but after trying a couple of examples you will find it a very efficient way of programming.

Line Length

Lines which are longer than the physical width of the page are wrapped around and right justified as in the GFA BASIC manuals. When typing in such lines you should ignore the wrap-around, and simply carry on typing the line to the end (this may require typing past the end of physical screen if the line is over 80 characters in length).

Remarks

Remarks are used, along with self-explanatory procedure and variable names, to help explain the actions of particular lines or groups of lines. These are always preceded by either REM or an exclamation mark (!). The ! type of remarks are the only type which GFA BASIC allows at the end of a program line. These will be used in a fashion similar to:

`Alert 2,"Ready To Print",0,"OK|Cancel",b !display print box`

The remarks are only included to aid understanding and may be omitted if this is desirable.

Screen Dumps

Except where specifically stated, all the screen dumps given in this book were produced from a high-resolution monochrome screen image using the Degas printer driver for nine pin Epson compatibles. This means that these dumps are of a 640x400 pixel resolution. However, some of the figures were produced on a 24 pin printer using specialised programs that I have written specifically for the task. These programs generally produce output with a very high resolution of 1416x1416 pixels.

Further information on the 24 pin technique is included in Appendix D, and details of more general screen dump methods are given in Appendix A.

Entering and Debugging the Examples

All the programs contained in this book have been thoroughly tested, and should hopefully contain no errors. The listings were all saved from GFA BASIC as ASCII files after being test run, and then directly imported into the text file. This means that little room was left for human error, and also means that indentation and variable capitalisation are shown exactly as they will appear when you enter them. However, errors may be induced while the programs are being typed in, and although some problems such as syntax errors and incomplete loops will be caught by GFA BASIC before the program is executed, other bugs are bound to creep through.

If you enter a program and it fails to produce the expected output, the first place to look is at lines containing numerical calculations, as a single wrong digit or variable can fundamentally alter the program's output. If, on running the program, you receive a *Procedure not found* error message, you have probably forgotten to include the relevant procedure from the library in Appendix A, as discussed above.

If you find typing too tedious and error prone I recommend that you obtain the support disk mentioned a few pages earlier, as this contains all the examples from this book in a form ready to run immediately from GFA BASIC.

2

Order and Chaos are Related

In general it is safe to say that a random process produces random results, and an ordered process produces ordered results. For example, randomly sprinkling something over a flat surface produces an even spread of particles with no structured pattern, but a set of explicit commands or equations (a deterministic process) gives a predictable, ordered result as in a computer program. There are, however, some processes which do not abide by this rule of thumb. The Sierpiñski triangle is the result of one such process.

The Sierpiñski Triangle

Because it was discovered by mathematicians in the early 1900s, the Sierpiñski triangle is one of the oldest fractals covered in this book. The reason for the triangle's

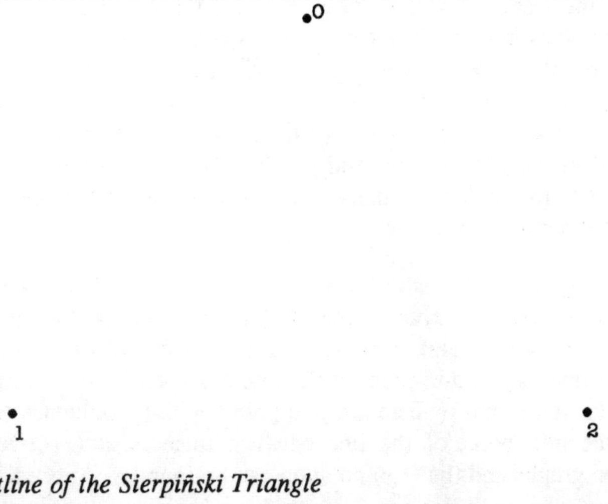

•0

•   •
1   2

Figure 2.1 Outline of the Sierpiñski Triangle

relatively early discovery is that it is based on very simple rules and can be drawn by hand, with little or no calculation. This meant that its creators were not hindered by their lack of computational machinery, the like of which is now indispensable in chaos research. Although the Sierpiñski triangle can be drawn by hand it does take a considerable amount of time, so to demonstrate it here we will be using the ST. However, if you would like to follow the original method you are quite welcome!

The starting point for the triangle is a blank plane, on which a triangle is outlined by three single points, numbered 0, 1 and 2 for reference. Such a situation is shown in Figure 2.1. The Sierpiñski triangle can now be created on this plane by applying the following rules:

1. Pick one of the vertices (corners) at random and go directly to it

2. Choose another vertex at random, move halfway towards it and plot a point

3. Repeat from step 2.

This type of process is known as an iterative process, because a single set of rules is applied repeatedly in order to produce the result. Each application of the rules is called an iteration, in our example the plotting of one pixel represents one iteration. Figures 2.2 shows the evolution of the triangle after two iterations.

It is quite natural to expect this random process to yield a random result but, as the lower part of Figure 2.2 shows, the process eventually gives rise to a rather complex, ordered structure - the Sierpiñski triangle. This unusual behaviour is best demonstrated with the aid of a simple program which executes the rules given above. Such a program is shown in Listing 2.1.

Basically the listing can be broken down into three sections. The first dimensions two arrays to store the x and y positions of each of the three vertices of the triangle, and places suitable values in them. The next section corresponds to stage one of the above process, as it initialises the drawing position by randomly setting it to be at one of three vertices. The drawing position can be thought of as the position of the computers 'pen' to give consistency with the manual drawing method. The drawing position is held in the variables px and py. The final section of the program is the REPEAT...UNTIL loop which actually does the drawing of the points, representing stages two and three of the process.

The only relatively difficult area in this program is the part which moves the graphics cursor halfway towards a vertex. This is done by finding the mid-point of the imaginary line between the current position and the relevant vertex, and moving to it. The method of finding the mid-point of the line has been borrowed from co-ordinate geometry, which states that if there are two points with co-ordinates (x,y) and (x_1,y_1) respectively, the mid point of the line between them is at $(x+(x_1-x)/2, y+(y_2-y)/2)$. More details on graphs and the (x,y) notation can be found in Appendix B.

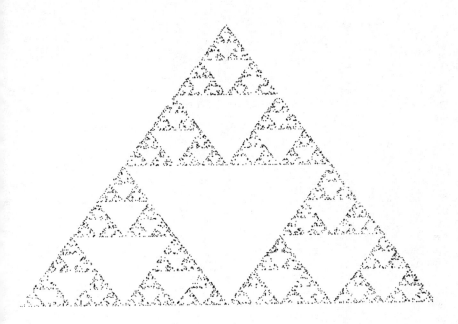

Figures 2.2: two iterations of the Sierpiński triangle

```
@Check_res !Get screen mode
'
Rem Dimension arrays to hold the vertex positions
Dim X(2)
Dim Y(2)
'
Rem Set positions of the three vertices X(0)=160
Y(0)=10
X(1)=25
Y(1)=190
X(2)=295
Y(2)=190
'
Hidem
'
Rem Pick a vertex at random
Vertex=Int(Rnd*3)
Px=X(Vertex)
Py=Y(Vertex)
'
Repeat
Vertex=Int(Rnd*3) !Pick vertex at random
Px=Px+(X(Vertex)-Px)/2 !Move to mid point of vertex
Py=Py+(Y(Vertex)-Py)/2 !held in vertex variable
@Fplot(PX,PY)
Until Mousek>0
```

Listing 2.1: This produces the Sierpiński triangle

When typing in Listing 2.1 you should ensure that you follow the guide-lines given in Chapter 1. However, as this is the first program in the book, explicit instructions will be given here to help you get the triangle up and running. First you should load GFA BASIC as normal and ensure that there are no programs in memory. Listing 2.1 should then be typed in exactly as shown (remarks may be omitted if desired).

Assuming you have corrected any syntax errors you may have made the program can now be executed by clicking with the mouse on the RUN option of the menu bar, or by pressing <SHIFT> and <F10>. If you have typed in the program correctly you will be presented with an error box displaying the *Procedure not found* message. At this point you should click on the Return button which will take you back to the editing screen, where the cursor will be placed at the @Check_res line. The reason for this error is that BASIC is trying to call the check_res procedure, but it can't find it. This is because the definition of check_res has not been typed in, it is one of those regularly used procedures which are listed in Appendix A.

This procedure definition should now be typed in at the end of the program. As this is the first program of the book, the check_res procedure is shown below to save flipping to the Appendix. A description of what this procedure does and how it works is given in Appendix A. When typing in this definition pay attention to the alert box line, because although it is shown on two separate lines for convenience it should be entered into BASIC as a single line, beginning with `Alert` and ending in the word `Dummy`.

```
Procedure Check_res
Res%=Xbios(4)
If Res%=1 !Medium resolution not supported
Alert 3,"This program only|works in high or|low resolution modes.",1,"                                                     Abort ",Dummy
Edit !Quit back to editor
Else
Monitor=Res%/2+1 !Set monitor variable
Endif
Return
```

The definition of the `fplot` procedure is also needed, and should be typed in as follows:

```
Procedure Fplot(X,Y)
Xp=X*Monitor
Yp=Y*Monitor
Plot Int(Xp)+Int(Frac(Xp)*2),Int(Yp)+Int(Frac(Yp)*2)
Return
```

If you now try to run the program you should see the triangle gradually appear, filling most of the screen. However, if you are using BASIC in medium resolution mode you will be presented with an alert box reminding you that the program does not work in this resolution. None of the programs do.

A Fractal

The Sierpiński triangle exhibits two important features which distinguish it as a fractal. The first is the fact that an immensely complex and structured pattern is created by just a few very simple rules. The Mandelbrot set is probably the most famous example of this, as one equation gives rise to an infinitely rich structure.

The second point is that if any part of triangular structure is sufficiently magnified the same general shape can be seen (see Figure 2.3). In this case the shape is an equilateral triangle. This property is called self-similarity, which also makes its most notable appearance in the Mandelbrot set. Note that the image cannot be optically magnified, as the resolution of the program's output is not good enough, and even if it was possible to produce more detailed output the paper can only be magnified to a certain level. This means that the image must be magnified mathematically by completely recalculating the part of the image that needs to be enlarged.

Unlike fractals, normal geometrical shapes are not self-similar and lose their identity when magnified enough, for example a circle becomes a straight line. An everyday example of a circle losing its identity is the phenomenon of the Earth appearing to be flat to a person standing on it, because they can only see a tiny fraction of its surface.

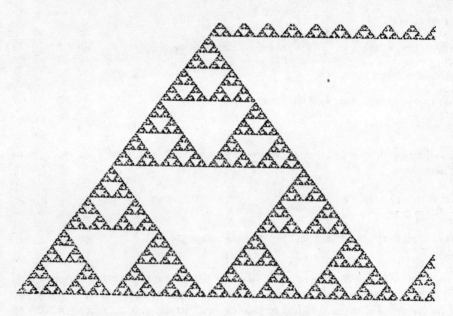

Figure 2.3: A magnified section of the Sierpiński triangle

Self-similarity can easily be seen in the Sierpiński triangle by altering the program in Listing 2.1. The easiest way to do this is to enlarge the triangle so that only a small part of it lies within the screen area, as shown in Figure 2.4. Using this method the program will still effectively move to each point, but will only plot points which lie on the screen. In practice the triangle is enlarged by altering the positions of the three vertices, which involves changing the section of the program which sets the vertex positions to read as follows:

```
X(0)=565
Y(0)=-530
X(1)=25
Y(1)=190
X(2)=1105
Y(2)=190
```

When enlarging fractals in this way it is imperative that the ratio of width to height (the aspect ratio) is kept constant. If this ratio is not maintained the fractal will become distorted, making self-similarity hard to identify. The Sierpiński triangle used in this chapter is 50% wider than it is tall and I have ensured that the enlarged triangle also has this property, thereby preserving the aspect ratio.

Something to note while experimenting with magnifications of the Sierpiński triangle, is its lack of substance. This may sound strange but in fact all the areas which appear to consist of solid black lines are actually full of self-similar triangular shaped holes. This continues to be true for consecutive magnifications, which suggests that the lines between the triangles are infinitely thin (hence non-existent), and implies the Sierpiński triangle to be nothing more than a group of holes! Many other objects with similar properties were devised at the same time, most notably the cube shaped Menger sponge, which, because it didn't really exist, could hold a volume of water equal to 100% of its volume.

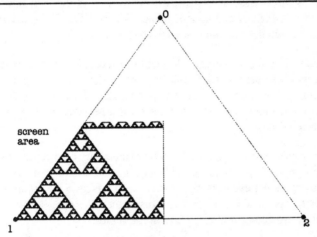

Figure 2.4: Magnification of the Sierpiński triangle by enlargement

Although turn of the century theorists were intrigued by the strange objects that they had discovered, they found it hard to continue their investigations due to the vast number of calculations required. Unlike most classic mathematics the outcome of these types of processes could only be determined by brute force calculation, and were very sensitive to small errors. Because of these problems objects such as the Sierpiński triangle were regarded as nothing more than mathematical oddities for many years, until the advent of chaos, when natural processes were shown to exhibit similar behaviour.

Population Dynamics and the Feigenbaum Diagram

Physicists and mathematicians are lucky. The problems they have to solve are generally based on inanimate objects which can usually be studied, or at least modelled, in a controlled laboratory experiment or on paper. Biologists, on the other hand, often have to study in a much larger laboratory of the Earth, where their experiments are constantly at risk from natural sabotage. The complex interactions between living organisms and their surroundings make it impossible to enact biological studies in the lab without upsetting the results, which means that patterns and predictions are very difficult to establish.

However, to make their lives somewhat easier, biologists have developed a branch of their science called ecology, in which individual creatures and events are generalised to produce simple mathematical equations which can predict the behaviour of large groups of organisms over a period of time. One such equation, used to predict simple population changes, is the May equation, developed in 1845 by Verhulst but much studied by the 'mathematical biologist', Robert May. It is this equation which forms the basis of the Feigenbaum diagram, named after Mitchell Feigenbaum, the first person to draw the whole diagram and analyse it in detail.

The Feigenbaum diagram represents the opposite situation to that of the Sierpiñski triangle. That is to say the complex, and in some places chaotic structure (see Figure 2.10) is generated from the very simple, non-random formula which May derived from population dynamics. The accepted name for a non-random process such as this is a deterministic process.

An example derivation of the equation can be obtained by considering the spread of a virus through a group of people. A virus has been used for the demonstration because viruses are among the simplest living organisms. In this case the population is defined as the number of people infected by the virus at any one time, and we are trying to find a method of predicting it at any stage in the future.

The accepted method of deriving a formula is to start from a very basic equation and then elaborate on each part of it in turn using other equations and basic facts. The equation we shall use as a starting point is shown below. It allows us to calculate the number of infected people at the end of any one week. The equation basically states that the number of people infected at the end of the current week (p_{new}) is equal to the number of people infected at the end of the previous week (p) plus the number of people who have been infected since then (n).

$p_{new} = p + n$

This equation is technically correct, but to calculate the population for any given week it is necessary to know the number of people infected during that week. This means that the equation can only be used to work out populations of weeks gone by, and cannot predict the future.

However, the need to know the number of newly infected people (n) can be eliminated because it can be shown that n is directly proportional to the number of people already infected at the end of the previous week (p). This is due to that fact that as the number of people infected increases so does the likelihood of an uninfected person coming into contact with an infected one, which in turn increases the chances of the healthy person being infected.

$$n \propto p$$

Of course, infected people cannot be re-infected so, assuming that all unwell people are contagious, the number of newly infected people must also be proportional to the number of healthy people. If, for example, everyone in the group was infected nobody else would be able to become infected, so n would be zero. Assuming that there are no in-betweens the number of healthy people must be equal to the total number of people in the group less the ones who are infected.

To eliminate the need to know the actual number of people in the group (and to show that this is unimportant to the equation) we can use percentages instead. In mathematics the numbers from 0 to 1 inclusive are used to represent percentages from 0% to 100%, where 0.5 is 50% for example (Appendix B gives more details on this and other pure maths topics). From now on we assume that p and p_{new} represent the percentage of the group infected at different times, and that n represents the percentage of the group who become infected during the week. The percentage of healthy people in the group is therefore 100%-p, as shown in Figure 2.5.

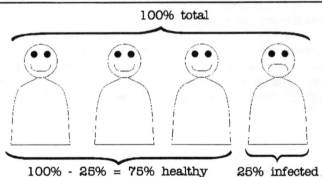

Figure 2.5: Pictorial representation of the method used to calculate the percentage of healthy people in the group

This can be represented mathematically as (1-p), remembering that 1 represents 100%:

$$n \propto (1-p)$$

Because n is directly proportional to both p and (1-p) we can say that n is proportional to $p*(1-p)$. An equation combining these facts is:

$$n = p*(1-p)$$

You may be able to guess that if this definition of n were substituted into the equation for p shown above, p would steadily rise to 100% and then stay there. If this were the case in the real world, and the group of people were the earth's population then the existence of the human race would be in doubt. In reality, however, certain people may not meet as many people as others, and a meeting does not mean definite infection. Also people may recover from the virus, either using the body's immune system, or with the aid of medical treatment (depending on the nature of the virus). For this reason a constant, c, must be included in the equation. This represents the success rate at which the virus spreads and persists in the human body, and will hereafter be referred to as the contamination constant.

Because n is proportional to c, we now have the following definition of n:

$$n = c*p*(1-p)$$

The result of substituting this definition into the equation for p is the May equation:

$$p_{new} = p + c*p*(1-p)$$

The beauty of this equation is its universality. As already discussed the use of percentages makes it independent of the number of people in the group. Since the number of people in any particular group is certain to change, due to births and deaths, this is of fundamental importance.

The introduction of the constant, c, also makes the equation more versatile by allowing it to be customised for any combination of virus and host. This means that the hosts do not have to be people, they could just as easily be animals or even computer disks! For a different virus/host combination a different value of c is all that is required.

Applying the Equation

Now that we have our equation we can use it to predict the likely number of infected people at the end of any week. Let us create a theoretical starting situation (call it week 0) in which 30% of the group have the virus ($p=0.3$) and the virus has a contamination constant of 1.9 ($c=1.9$). The calculations to determine the number of people infected after the first four weeks are shown below.

Time	Calculation	People infected
Week 0	$p = 0.3$	30%
Week 1	$p = 0.3 + 1.9 * 0.3 * (1 - 0.3) = 0.7$	70%
Week 2	$p = 0.7 + 1.9 * 0.7 * (1 - 0.7) = 1.1$	110%
Week 3	$p = 1.1 + 1.9 * 1.1 * (1 - 1.1) = 0.89$	89%

The results are calculated by feeding the previous week's population into the next week's equation. This equation is therefore said to be dependant on mathematical feedback, the flow of which is shown in Figure 2.6. Feedback occurs in many real world situations. For example audio feedback is induced in a PA system if the microphone is brought too close to the amplifier.

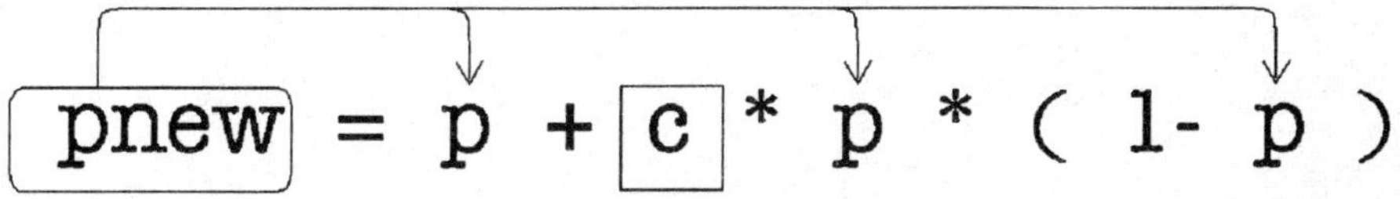

$$\boxed{pnew} = p + \boxed{c} * p * (1- p)$$

Figure 2.6: Mathematical feedback in the May equation

The Sierpiński triangle also relies on feedback. Each time the rules are applied their starting point is taken from the previous iteration. Like the Sierpiński triangle, our population simulation is an iterative process, but in this case an iteration is defined as one application of the formula. Such processes are notoriously difficult to predict, as there is no way of telling what a particular set of initial conditions will lead to without working through the whole process. Unfortunately many natural things like the weather are subject to feedback, and this is why powerful super-computers are needed to calculate forecasts. The more formal name used to describe such a process is 'non-linear dynamic system', and it is this definition which will be found in the more mathematically minded chaos works.

From the time series graph of these results (see Figure 2.7a) it can be seen that the population oscillates back and forth a few times before coming to rest at 100%. This behaviour is typical of a feedback dependant process. It begins in an unstable state and gradually returns to its equilibrium position, where it is stable. A real life example of this effect in action can be demonstrated by disturbing a simple pendulum, it will oscillate with a decreasing frequency until it hangs vertically in equilibrium. The graph also reveals a flaw in our formula - during the first few weeks the number of people infected actually exceeds 100%. This is not actually an error because values above 100% can be taken to represent over-population, where there are more occurrences of the virus than there are people.

As we already know, populations below 100% occur when some people remain uninfected by the virus, in other words when the virus is in a state of under-population. During a period of under-population the virus generally thrives, spreading itself to as many healthy people as possible. But when the people become over-populated many of the viruses cannot sustain reasonable living conditions and therefore die. This results in the population always being attracted to the best possible population value, 100%, and for this reason $p=1$ is said to be the attractor of the May equation (this is only true for values of c within a certain range).

As previously mentioned a different value of c can be used to represent a different virus. Figure 2.7(b) shows graphically the population of the virus with c=2.3. Again, there is some oscillation in the first few weeks before the virus settles down, but unlike the previous example this virus settles down into a two point oscillation rather than a single value.

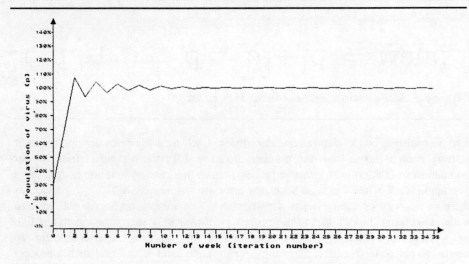

Figure 2.7 (a): Time series graph for the May equation where c = 1.9

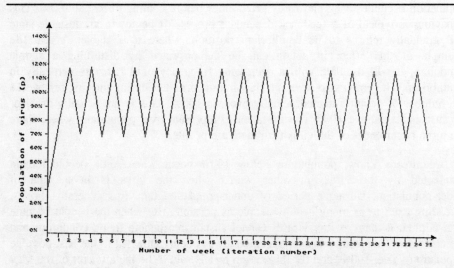

Figure 2.7 (b): Time series graph for the May equation where c = 2.3

Order and Chaos are Related

Further values for c give even more interesting oscillations. For instance when $c=2.5$ the population settles down to oscillate between four values (see Figure 2.7(c)). By the time viruses such as that with a contamination constant of 2.9 are reached the population graphs have degenerated into chaos (see Figure 2.7(d)), with p jumping between many different values with no apparent rhythm.

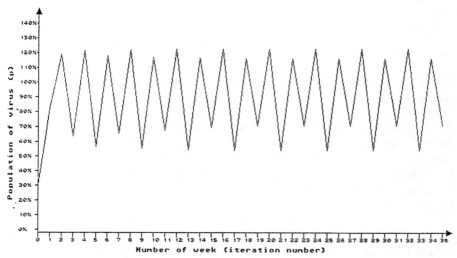

Figure 2.7 (c): Time series graph for the May equation where $c = 2.5$

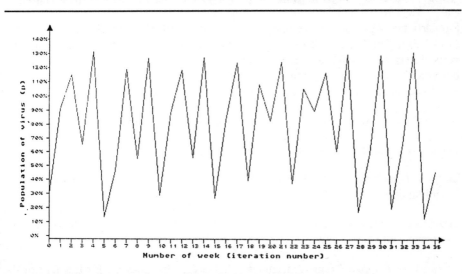

Figure 2.7 (d): Time series graph for the May equation where $c = 2.9$

A simple graph plotting program can be used to experiment with various viruses (i.e. values of *c*). Such a program is shown in Listing 2.2.

```
@Check_res
Input "Contamination constant, c:",C
P=0.3
@Fplot(0,200-P*140)
For Week=1 To 80
P=P+C*P*(1-P)
@Fdraw(Week*4,200-P*140)
Next Week
@Waitmouse
```

Listing 2.2: A simple program used to produce a time series graph from the May equation

This program requests a value for the contamination constant, *c*, and then performs the May equation for this value over a period of 80 weeks, drawing a time series graph (with time plotted horizontally and population vertically) as it goes. To keep the listing short no axes are plotted, but this does not matter because the actual values are unimportant, the program is simply used to demonstrate different behaviour patterns.

The main body of the program is the FOR...NEXT loop containing the easily recognisable May equation, and the `fdraw` command which draws a line from one point to the next. Note that because the range of the week variable (1 to 80) is small in relation to the horizontal screen resolution (0 to 319) used by `fdraw`, the *x* position passed to `fdraw` must be multiplied by 4 so that the whole screen is used.

Similarly the population variable, *p*, has to be made larger due to its very small range (0 to 1.3), by multiplying it by 140. However, due to the inverted nature of the ST's vertical axis the amplified value of *p* must be taken away from the maximum vertical position (200) to ensure that the graph is plotted the right way up. More details on the ST's co-ordinate axes are given in the graphs section of Appendix B.

Experimentation with this program should establish the following facts:

❏ *p* always takes a few weeks to settle into a pattern

❏ In most cases the higher the value of *c* the more values *p* oscillates between after settling down

❏ *p* always oscillates between an even number of values, except in the chaotic regions, and when *c* is one of a certain set of values (try *c*=2.83)

❏ Values of *c* above 3.0 give meaningless results, as the attractor for this part of the diagram is minus infinity.

Using this program it would be very time consuming, and difficult, to find the exact value of c at which the number of values oscillated between curiously goes from 1 to 2. A better way to discover this and other things is to combine the graphs of all the possible values of c in one graphical representation. The composite graph that results from doing this is referred to as the Feigenbaum diagram, the generation of which is a relatively simple affair, now that we have a program to calculate successive values of p.

To draw the Feigenbaum diagram it is necessary to compress the time series graph for each value of c so that it fits into one vertical column of pixels on the screen (i.e. so that it is one dimensional). All these compressed graphs are then drawn horizontally across the screen, so that they join together to form a two dimensional full-screen map of p against c. This process is shown pictorially in Figure 2.8.

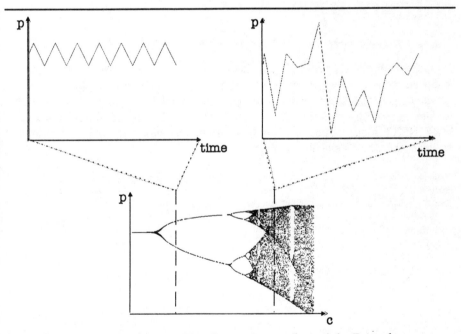

Figure 2.8: Relationship between May time series graphs and the Feigenbaum diagram

The first stage of creating the compressed graph can be done by adapting the time series program to the form shown in Listing 2.3. There are two important changes here. Firstly the virus population is now calculated for 100 weeks, but the first 50 of these calculations are ignored to ensure that p is only plotted after stabilisation. Because the system relies of feedback the first 50 calculations must be performed, even though they are not used.

Secondly, all points on the graph are now plotted at the same horizontal position, and are not connected by lines, making oscillatory patterns easily identifiable. You could try experimenting with this program using the notable values of c mentioned above to gain an understanding of the program's function.

```
@Check_res
Hidem
'
Input "Contamination constant, c:",C
P=0.3
@Fplot(0,200-P*140)
For Week=1 To 100
P=P+C*P*(1-P)
If Week>50
@Fplot(160,200-P*140)
Endif
Next Week
@Waitmouse
```

Listing 2.3: The compressed graph plotting program

If graphs of this sort are now plotted side by side for values of c between 1.8 and 3.0 inclusive (values below 1.8 are uninteresting, as they stabilise at a single value) we get the Feigenbaum diagram shown in Figure 2.9 (overleaf). Listing 2.3 can be altered for this purpose by replacing the human input for c by one produced by a BASIC FOR...NEXT loop. Listing 2.4 has had this modification included so that it now produces the whole Feigenbaum diagram.

```
@Check_res
Hidem
For C=1.8 To 3 Step 0.00385/Monitor
P=0.3
For Week=1 To 100
P=P+C*P*(1-P)
If Week>50
@Fplot((C-1.8)*260,200-P*140)
Endif
Next Week
Next C
@Waitmouse
```

Listing 2.4: This generates the complete Feigenbaum diagram

The Feigenbaum diagram is not one of the most aesthetically pleasing fractals, but it is a stunning illustration of how a simple, non random, iterative process can produce a finely structured image in some places, and total chaos in others.

One result of much scientific study of the diagram is that conventions have been created to describe some of its most important features. The section of seemingly random pixels to the right of the diagram is referred to as the chaotic region. Note that this area is not actually random, as the results are the same each time the program is executed.

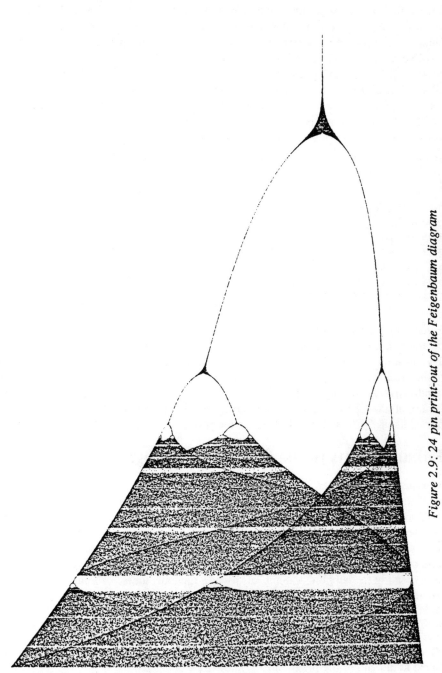

Figure 2.9: 24 pin print-out of the Feigenbaum diagram

The point where a single line splits into two is called a bifurcation and the splitting, or bifurcating, which occurs at these points is referred to as period doubling, as the number of equilibrium states (the period) doubles. The period is the number of possible values that p oscillates between after stabilising, for example the section between 2.44 and 2.54 is period four. By the very nature of period doubling almost all sections past $c=1.95$ have an even number of possible states, but there are however some 'windows' of order occurring in the chaotic regions which are of period three. Table 2.1 shows the period for various sections of the Feigenbaum diagram.

Table 2.1: The number of equilibrium states for selected sections of the Feigenbaum diagram

Range of c (approx)	Period
0.00 to 1.95	1
1.95 to 2.44	2
2.44 to 2.54	4
2.54 to 2.56	8
2.83 to 2.84	3

Figure 2.10 (overleaf) shows a section of the main window in the chaotic region enlarged to about seven times its original size. This reveals some interesting features. The most striking of these is the very noticeable self-similarity. Inside the window a miniature Feigenbaum can be found, complete with characteristic period doubling and window-ridden chaotic region. This section also highlights the incredibly random nature of the chaotic region. Comparing this to a screen full of pixels randomly placed using GFA BASIC's RND function reveals a striking similarity.

Enhancements to the Feigenbaum Program

Musical fractals are rare, but the Feigenbaum diagram is one of the few that lends itself to the addition of sound. By playing a tone proportional in pitch to the value of the population, p, whenever a point is plotted the sound can be used to give an audio representation of how the Feigenbaum diagram degenerates into chaos. At the left hand side of the diagram the period is 1, so the tone is constant. After bifurcation the tone oscillates between two pitches, and by the end of the diagram chaos has broken out, creating a bubbling sound of pseudo-random tones. Adding the following command after the `fplot` line will create the sound:

```
Sound 1,15,#(100+P*760),0
```

Because sound in GFA BASIC works under interrupts the line below will need to be added at the end of the program to stop the tone being produced.

```
Sound 1,0,#0,0
```

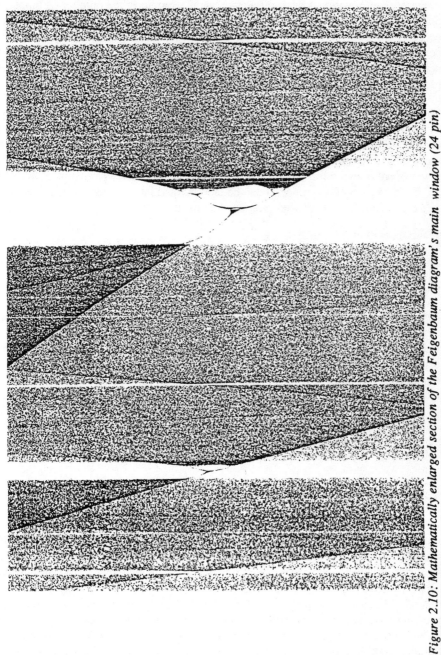

Figure 2.10: Mathematically enlarged section of the Feigenbaum diagram's main window (24 pin)

A piece of information lost in the Feigenbaum program is the number of times each point has been plotted, which would be useful when looking at the number of equilibrium states in different parts of the Feigenbaum diagram. For example, in the period one section of the diagram each point is plotted 50 times, whereas in the period two region every point is plotted 25 times, and so on until in the chaotic region most points are not plotted more that once.

The number of iterations for each point could be conveyed either by using colour or by plotting the diagram in three dimensions, where the third dimension represents the number of times the point is plotted. The three dimensional 'Feigenbaum landscape' method is discussed in Chapter 7, as it obviously requires a knowledge of three dimensional drawing techniques. The colour method is somewhat simpler and, if you are using low resolution mode, it can be incorporated into Listing 2.4 by replacing the fplot line with the following program fragment:

```
Xp=(C-1.8)*260 !Calculate x and...
Yp=200-P*140 !...y position of point
Oldcolour=Point(Xp*monitor,Yp*monitor) !Determine old colour
Color (Oldcolour+1) Mod 16 !Set new colour
@Fplot(Xp,Yp) !Plot the point
```

Note the way that the x and y position of the point are calculated and stored to save having to calculate them twice later on. The actual colour in which a point is plotted is determined by finding the old colour of the point at the plotting position (found using the POINT function) and adding 1. Because the ST only has 16 colours, modulus 16 is taken of the colour value in order to keep the colour within the relevant 0 to 15 range (for more information about MOD see Appendix B).

At first this process may sound like the start of a magic trick, but it is actually the simplest way of changing the colour of a pixel each time it is plotted so that the colour of the point shows how many times it has been visited.

Conclusion

In this chapter it has been shown, using the Feigenbaum diagram, that a simple mathematical equation, with no random elements can give rise to what appears to be a totally random result, and that a random set of rules can give rise to a pattern of very rich structure, such as the Sierpiñksi triangle. This two way relationship between order and chaos is more than anecdotal, it represents a whole new way of perceiving natural processes occurring around us. For although our simple virus population program has inaccuracies it still illustrates the important point that a natural process such as population growth and decline can be described in one simple equation with nothing more than some common sense and a little help from the laws of statistics. Is it possible, then, that even the most complex natural processes can be reduced to simple equations, thus allowing them to be predicted with relative ease? This important question is dealt with in the next chapter.

3

Weather, Chemistry and Strange Attractors

Weather forecasters have a bad reputation. People find it incredible that a forecaster can proudly announce *"there will be no hurricane"* merely hours before the country is hit by some of the worst storms in living memory. In Britain forecasters are particularly unfortunate because our island location means that they have to deal with a wide variety of weather, from blizzards to heat waves.

The seeds of modern forecasting methods were sewn in the 50s and 60s, with the advent of computers. Meteorologists had long believed that they could represent the world's weather using a complex mathematical model, and now hoped that computers would be able to perform the multitudinous calculations needed to run such models. With the advantages of accurate weather forecasting being obvious to even the most unscientific of people much time and money was put into meteorology research. This was spent on various projects, ranging from collecting weather data to building specialised weather forecasting computers. For decades the most powerful civilian computers were always to be found at weather forecasting centres.

One beneficiary of the forecasting boom was Edward Lorenz, a research programmer at the Massachusetts Institute of Technology. Being a keen mathematician and an experienced meteorologist Lorenz was quite at home writing programs on his rather elementary computer (by today's standards) to model simple weather systems. In his first model Lorenz simplified the weather down to a fluid dynamics system, sea and air being fluids, and chose 12 relevant equations each representing a different element of the weather, and each interacting with the others.

The model was very simple, some thought too simple, but the output from it had some interesting similarities with the behaviour of real weather. Everything that happened in the Lorenz model followed on from what had happened previously. It was by no means random, but the exact same series of events never happened twice. Lorenz managed to interest many colleagues in the model, and despite the fact that all it did was continuously churn out numbers on a teletype (he had no graphics facilities) they would enjoy visiting Lorenz's office to see what the simulated weather was up to.

Because the program was rather slow Lorenz decided to try and simplify his model further, as far as possible without destroying its realistic unpredictability. He eventually trimmed the model down to the three non-linear differential equations:

$dx/dt = a*(y-x)$

$dy/dt = b*x-y-x*z$

$dz/dt = x*y-c*z$

where:

❏ x, y and z are variables representing different weather aspects;

❏ t is time;

❏ a, b and c are constants.

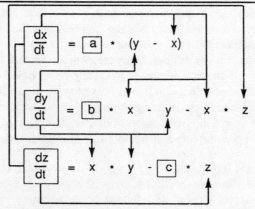

Figure 3.1: Feedback in the Lorenz model

The way that these equations were derived places their detailed explanation beyond the scope of this book, but knowledge of how to use them is important for later examples. As with the May equation described in the previous chapter, the non-linearity of the equations is due to the mathematical feedback created after successive iterations. The feedback in the Lorenz model is a good deal more complex than in the population simulation because there are three equations and three variables, as shown in Figure 3.1.

The Lorenz equations are referred to as differential equations because, unlike normal ones, they are used to find the rate of change of variables, rather than the absolute values. For example dx/dt represents the rate of change of x, or more explicitly the (small) change in x that occurs in the (small) time, t. The d in front of the variables stipulates that the changes in them must be small - technically the d should be written as δ (the Greek letter delta). The reason that the change must be small is that the equations were not designed for large changes, to perform a large change you have to carry out the equivalent group of small ones.

The actual values of x, y and z are calculated by repeatedly adding the change in the variable (e.g. dx) calculated using the formula, to the previous value of the variable (e.g. x). Lorenz's original programs used these equations to simply produce a list of numbers, similar to those shown below:

iteration:0	x=1.000	y=1.000	z=1.000
iteration:1	x=1.000	y=1.260	z=0.983
iteration:2	x=1.026	y=1.518	z=0.970
iteration:3	x=1.075	y=1.780	z=0.959
iteration:4	x=1.146	y=2.053	z=0.953
iteration:5	x=1.236	y=2.342	z=0.951

These figures meant a lot if you could decipher them, but general trends and repetition were hard to spot. However, Lorenz invented a primitive kind of graphics on his system to make the results more accessible. He chose one of the variables and instead of plotting it numerically he scaled it so as to be in the range of 0 to 79 and then plotted an 'a' character on the current printer row, that number of spaces across the page. Over a period of a few days a rather long time series graph would be produced showing the strange non-periodic behaviour of the model. Similar output is shown in Figure 3.2 (overleaf).

Graphics are now almost a standard feature of modern microcomputers, and an acceptable line graph of the information can easily be created on the ST with the aid of a simple program, such as the one given in Listing 3.1.

This program can be broken down into two main sections. The first is the initialisation of all the constants, variables and graphics facilities. The second is the loop in which successive values of x, y and z are calculated, and where z is plotted vertically against time. It is important that all three variables are calculated, even though only one is displayed, because all the variables interact.

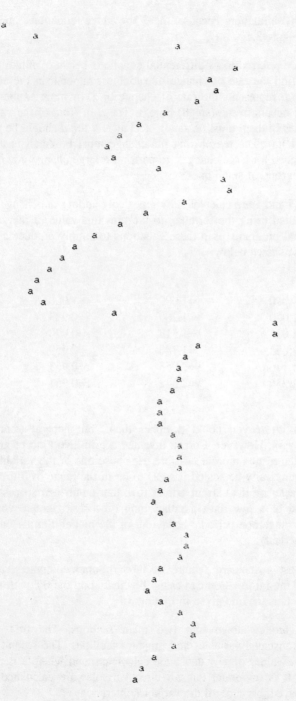

Figure 3.2: Simulation of Lorenz's first time series graphs

Weather, Chemistry and Strange Attractors

```
@Check_res
'
Rem Set up constants
A=10
B=28
C=8/3
Dt=0.01
Detail=10
'
Rem Set up initial values of variables
X=1
Y=1
Z=1
'
Rem Move to starting point and hide mouse pointer
@Fplot(-1,0)
Hidem
'
For T=0 To 319
For Dummy=1 To Detail
'
Dx=A*(Y-X)  !Calculate increments
Dy=B*X-Y-X*Z !
Dz=X*Y-C*Z  !
'
X=X+Dx*Dt !Calculate new values for x, y and z
Y=Y+Dy*Dt !
Z=Z+Dz*Dt !
Next Dummy
'
@Fdraw(T,200-Z*3.5) !Plot the line
'
Next T
@Waitmouse
```

Listing 3.1: A program to produce a time series graph from the Lorenz equations.

There are several experimental alterations that can be made to this program, the most obvious being the values of the constants a, b and c, and the initial values of x, y and z. You could also change the fdraw call so that x or y are plotted instead of (or as well as) z. The detail of the plot may also be altered by using the detail variable, which determines how many 'dummy' iterations are executed for each point plotted on the screen - the larger the number the less the detail (but the more time you can fit on the screen).

As previously discussed, the dummy calculations must be processed even though they are not plotted, as large changes in the variables (and in time) cannot be made due to the nature of the equations. Note that if you do change this variable or the constants you will probably have to change the scaling factor used to calculate the y parameter of the fdraw call. However, this is not all that easy, as it is impossible to calculate the maximum and minimum values of x, y and z without actually calculating the result of the three equations for every iteration.

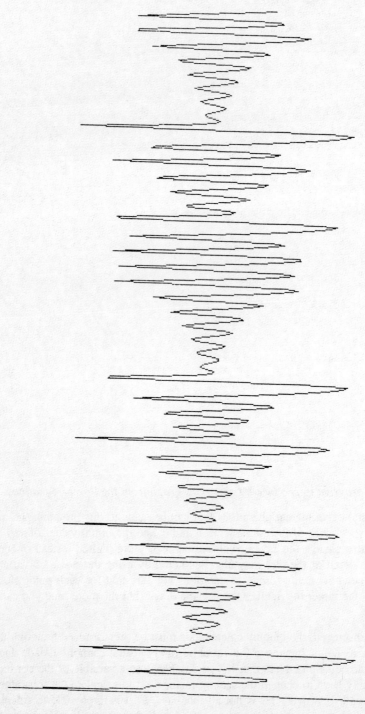

Figure 3.3: Time series graph produced by Listing 3.1

The kind of output that you can expect is shown in Figure 3.3. No axes are drawn as actual values are irrelevant without a full description of the equations' derivation.

The graphs produced by this program are not all that aesthetically pleasing, but they do allow the non-periodic repetition of similar events to be spotted with relative ease. An example of such an event is the vertical oscillation which gets larger and larger before collapsing and beginning again, with a slight difference.

The Butterfly Effect

The is no doubt that Lorenz's carefully planned experimental models were very enlightening, but probably his most important discovery was accidental. Wishing to hurriedly continue an earlier run of his model Lorenz used initial values of x, y and z taken from an earlier printout. He assumed the fact that the values on the printout were rounded off to three decimal places (from his computer's more accurate internal values) to save space would not make much difference to his readings.

To simulate the original run of the model we can alter the relevant lines in Listing 3.1 so that the initial values of x, y and z are set up as follows (it is important to type the numbers in full):

```
X=8.1646812788
Y=8.9676166531
Z=25.500286726
```

These numbers are of the same precision as those used internally in the program by GFA BASIC, and are said to be shown to 11 significant figures. If the program is now executed (with the `detail` variable set to 5) a graph similar to that shown in Figure 3.4(a) will be produced, which is not particularly unusual. However, if the program is executed with the initial values rounded off to four significant figures completely different results can be seen.

The initial values can be rounded off by changing the relevant lines to read as follows:

```
X=8.165
Y=8.968
Z=25.50
```

Rounding these numbers off is effectively what Lorenz did by typing in the old printed out values. The output from these less precise initial conditions is shown in Figure 3.4(b).

Comparing the two lines reveals that although they follow the same path for the first few iterations they gradually diverge into totally different patterns. Figure 3.4(c) illustrates this more concisely by super-imposing the two lines on the same graph. The effect witnessed here is what is technically known as sensitive dependence on initial conditions, meaning that the equations are so sensitive to the initial conditions supplied to them that even the smallest difference in the initial values quickly grows into a large difference.

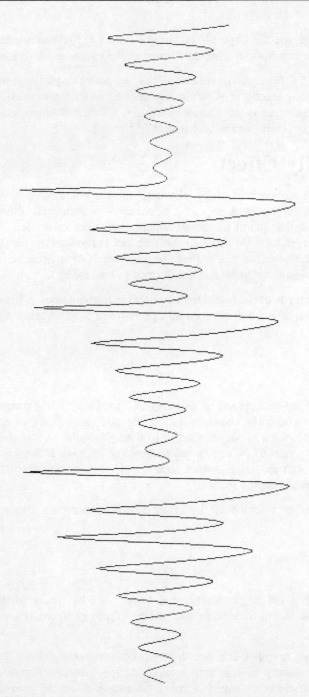

Figure 3.4 (a): Time series graph of initial conditions accurate to 11 significant figures

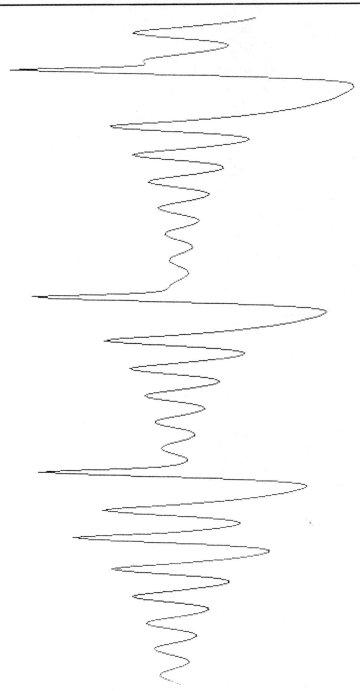

Figure 3.4 (b): Time series graph using initial conditions accurate to three significant figures

Figure 3.4 (c): Figures 3.4 (a) and 3.4 (b) superimposed

Lorenz immediately realised what his results meant. They showed that his model, and hence the weather, had a sensitive dependence on initial conditions. Ultimately this meant that weather prediction was practically impossible, because even the smallest error from a thermometer or other instrument would send a forecaster's model off down an increasingly wrong track. Other meteorologists took a more optimistic view, believing that they could alter future weather by making the relevant small changes such as heating small areas of land.

They were of course right, but Lorenz pointed out they could not control the weather with any precision, as the effects of the small changes they made would be impossible to predict because other, undetected, small changes would have equally large results. Some argued that a complex system such as the weather was so sensitive that the path of a hurricane could be reversed by a butterfly taking off from the other side of the globe. Hence the name 'butterfly effect' was coined as the common term for sensitive dependence on initial condition.

A more down-to-earth example of the butterfly effect is the story of an airline company going bankrupt. A man leaves his house on foot one morning, with the purpose of going to the travel agents to book his holiday. After walking a few yards down the street he looks down and finds that his shoelace has come undone. Naturally he bends down to tie the lace, but in doing so he loses his balance and falls into the road. As he stumbles to his feet a passing bus collides with him and he is subsequently taken to the local hospital. Because of this the man is unable to book his flight before the deadline, meaning that the airline company is one person short of the quota needed to make the flight financially viable. The flight is cancelled, and ticket holders refunded, but with the absence of the income from that flight the already troubled airline company gets into financial difficulties and goes bankrupt. It seems incredible, you never read of companies going bankrupt because of undone shoelaces, but it could happen, and it would be almost impossible to predict.

As the airline example shows, weather is not the only thing that would suffer because of the butterfly effect, world money markets, politics and the lives of individuals can be drastically affected by even the smallest changes. The airline discussed above could have been a major company, and its downfall could have pushed an already gloomy economy below a critical level and toppled the government and lead to civil war.

The butterfly effect raises important questions about the wisdom of organisations around the world who expend considerable resources on weather prediction. However, in the short term the errors are small, so forecasting a day or two in advance is just about possible with a fast computer, accurate algorithms and a bit of luck, but past that predicting the weather becomes more of an art that a precise science.

At first it seems strange that something as simple as the Lorenz model is so unpredictable, when more complex periodic systems such as the ebb and flow of the tides, and planetary motions are so easy to forecast. The reason that periodic systems are much easier to predict is that their repetitions are easy to observe and describe mathematically. Almost all non-linear dynamic systems (including the Lorenz model) are non-periodic, meaning that they are prone to the butterfly effect, and hence unpredictability and chaos.

The Lorenz Attractor

The Lorenz model has three variables, three directions in which it can change, and is therefore said to be three dimensional. This means that instead of simply plotting a time series graph of z against time we could plot x against y against z on three dimensional axes, like those shown in Figure 3.5.

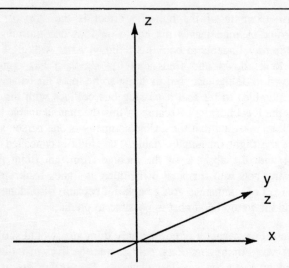

Figure 3.5: Typical three dimensional axes

The shape produced is known as the Lorenz attractor, an infinitely long complex spiral which never intersects itself. Plotting it in all three dimensions is rather involved and unnecessary, as a good representation can be achieved by plotting z against x in two dimensions.

Figure 3.6 shows the graph that results from this process, but to fully appreciate the line's behaviour it is necessary to see it being drawn. Listing 3.2 can be used for this purpose.

Weather, Chemistry and Strange Attractors

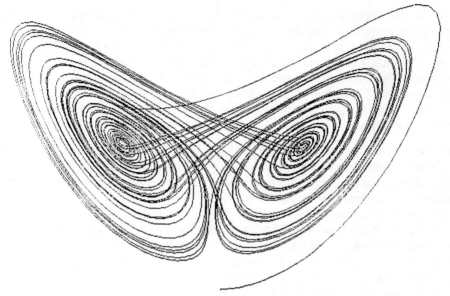

Figure 3.6: The Lorenz attractor for a=10, b=28, c=8/3

```
@Check_res
'
Rem Set up constants
A=10
B=28
C=8/3
Dt=0.01
'
Rem Set up initial values of variables
X=1
Y=1
Z=1
'
Rem Move to starting point and hide mouse pointer
@Fplot(157,197)
Hidem
'
Repeat
'
Dx=A*(Y-X) !Calculate new increments
Dy=B*X-Y-X*Z !
Dz=X*Y-C*Z !
'
X=X+Dt*Dx !Calculate new values for x, y and z
Y=Y+Dt*Dy !
Z=Z+Dt*Dz !
'
@Fdraw(150+X*7,200-Z*3.5) !Plot the line
'
Until Mousek>0
```

Listing 3.2: Plotting z against x in two dimensions to create the Lorenz attractor

The Lorenz attractor takes its name from the mathematical definition of such an object. An attractor is a set of points, in an imaginary mathematical space known as phase space. Each point on the attractor represents the state of the system at a particular time, and the line showing the sequence of these points is known as the trajectory. By looking at the trajectory of the attractor the behaviour of the system can be determined. The idea of attractors has been around for some time, but non-periodic attractors (called strange attractors) such as the one created from the Lorenz model are relatively new.

As this program shows, the trajectory behaves unpredictably, switching from lobe to lobe almost at random, but it does produce a very complex structure. One of the curiosities of the line is that it never intersects itself, although on the two dimensional screen of the computer it looks like it does. Because the Lorenz attractor is non-periodic it cannot intersect itself, if it did it would rejoin an old trajectory and periodically repeat. However, if the line never crosses and goes on forever it must be true that an infinitely long line can exist in a finite three dimensional space. As with many chaos ideas this obviously contradicts the laws of physics, but in fact it is possible if the space in which the line is enclosed is what is known as folding space (see the books listed in the Bibliography for details).

The Lorenz attractor is similar to many natural systems which exist in real space. For instance the turbulence in a fluid and the trajectory of an object orbiting a two planet system have similar properties. A recently developed executive toy, consisting of a metal pendulum oscillating over a bed of magnets, also exhibits similar behaviour.

The butterfly effect raises an interesting question: what is the real Lorenz attractor? Lorenz drew his first diagrams using output from a program using numbers correct to six decimal places. GFA BASIC has a precision of 10 decimal places, so it will produce a different output to that achieved by Lorenz. So who is right? This is a good question, but although the values may differ the general characteristics are the same. I was interested to see noticeably different behaviour when converting the GFA BASIC (10 decimal places) Lorenz attractor program to STOS BASIC (8 decimal places) during the preparation of this book.

The Rössler Attractor

Using three non-linear differential equations derived from chemistry it is possible to draw another strange attractor, the Rössler attractor. This exhibits all the same features as the Lorenz attractor, but is a different shape (see Figure 3.7). The equations are as follows:

$dx/dt = -(y + z)$

$dy/dt = x + y * a$

$dz/dt = b + z * (x-c)$

Listing 3.3 can be used to draw the attractor. Note that in most respects it is the same as the Lorenz program (Listing 3.2), except that the three equations are different and a pseudo-3D effect is created by plotting $y+2*z$ (instead of just z) against x.

```
@Check_res
'
Rem Set up constants
A=0.2
B=0.2
C=5.7
Dt=0.005
'
Rem Set up initial values of variable
X=-10
Y=-1
Z=-1
'
Rem Move to the starting point and hide mouse pointer
@Fplot(60,155)
Hidem
'
Repeat
'
Dx=-(Y+Z) !Calculate new increments
Dy=X+Y*A !
Dz=B+Z*(X-C) !
'
X=X+Dt*Dx !Calculate new values for x, y and z
Y=Y+Dt*Dy !
Z=Z+Dt*Dz !
'
@Fdraw(150+X*9,150-(Y+2*Z)*2) !Plot the line
'
Until Mousek>0
```

Listing 3.3: Drawing the Rössler attractor

Again the constants and initial variables can be altered to produce a different Rössler attractor.

Enhancements to the Attractor Programs

It has been mentioned in this chapter that the trajectory of a strange attractor never intersects itself. This is obviously true because the systems on which strange attractors are based are non-periodic (non-repeating), and for the trajectory to go through the same point twice would obviously be a repetition. Because the attractors are three dimensional but the ST's screen is only two dimensional it often looks as though intersections are occurring, where lines are actually passing behind or in front of one another.

The simplest way of verifying the non-periodic nature of the attractor is to use colour to represent the third dimension. If each point was plotted in a colour proportional to the value of y in the program given in Listing 3.2 then, ideally, two adjacent pixels would be touching only if they were both of the same colour. However, the range of values for y is considerably larger than the range of colours on the ST making this method unusable.

An alternative, and very impressive, way of getting round this problem is to write a program to produce an isometric 3D attractor which rotates around the z axis. The rotation gives the impression that the attractor is truly three dimensional, and just by watching it spin it can be seen that the line doesn't cross. Such a program is beyond the scope of this book, but a less ambitious one allowing static 3D versions of the attractor to be drawn at any angle of rotation would allow the same observations to be made. Three dimensional drawing methods are introduced in Chapter 7.

There are many more strange attractors which are just as interesting as those discussed here. Information on other attractors can be found in some of the books listed in the bibliography.

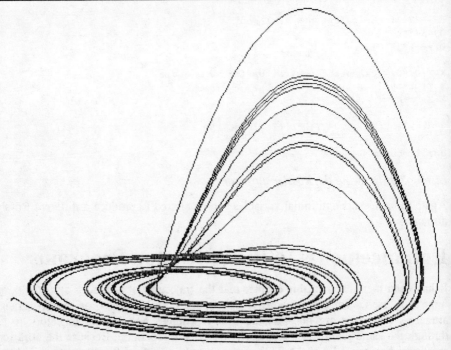

Figure 3.7: The Rössler attractor

The Mandelbrot Set

If you asked a scientist or computer user, of any nationality, what single word comes into their mind when chaos or fractals are mentioned they would most probably say 'Mandelbrot'. The colourful Mandelbrot set is by far the most famous image from the world of chaos, and has now become the emblem of the science. The set's fame has been brought about by its propagation throughout the world by the media, who picked up on this particular branch of modern science because it produces aesthetically pleasing pictures, in colour, which can be appreciated by everyone.

The set has now found its way into all manner of places, from T-shirts and record covers to newspaper articles about the economy. Unfortunately, however, the origins and meaning of the set are often distorted or even lost during the journey to the popular press. Even computer magazines only print programs to draw pretty pictures, rather than explain the underlying concepts. This chapter has been written to help rectify the situation, using various programs to demonstrate the theory.

Benoit Mandelbrot could not have anticipated that his work would have such wide popular appeal when he saw the set for the first time in early 1980. Mandelbrot's first plots were ill-defined monochrome printouts, interesting not for their appearance, but for their relevance to pure mathematics research. This research concerned the behaviour of iterative processes involving complex numbers. Something which often appears contradictory in chaos literature is that the process for plotting the Mandelbrot set is said to be very simple (just one equation), but complex numbers make the process seem a good deal more involved. In actual fact complex numbers are really just simple mathematical tools, and not a crucial part of the process at all.

It is interesting to note that the Mandelbrot set, unlike most of the fractals discussed so far, originated from purely mathematical beginnings, not from a study of nature. However, the set does exhibit the interesting features associated with other fractals, most notably the infinite complexity which results in the most spectacular self-similarity seen in this book.

This chapter describes two methods for drawing and experimenting with the Mandelbrot set. The first uses a simple scenario of a circle, a line, and some elementary maths. The second method follows Mandelbrot's original technique of using complex numbers. If you feel your maths is not up to the relative challenge of complex numbers you will probably find it easier to follow the circle method, but a subsequent glance at complex numbers may prove fruitful. Note that both methods result in the same program.

The Circle Method

Like the strange attractors discussed in the previous chapter the Mandelbrot set is based on a deterministic non-linear process, but whereas the results of the Lorenz and Rössler equations are drawn directly onto the screen as a graph the relationship between the Mandelbrot equation and the resulting set is more complicated.

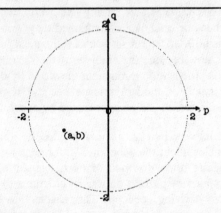

Figure 4.1: The Mandelbrot Circle

The first part of the process takes place in a circle such as that shown in Figure 4.1, with centre (0,0) and radius 2. Inside this circle there is a single point, at an initial position (a,b), to which the following non-linear equations can be applied (for details of indices and the ^ notation see Appendix B):

$p_{new} = p^{\wedge}2 - q^{\wedge}2 + a$

$q_{new} = 2*p*q + b$

In GFA BASIC this is most suitably written:

```
pnew = p*p - q*q + a
qnew = 2*p*q + b
```

Figure 4.2 helps to explain further these two equations by showing the feedback that occurs between them on successive iterations.

The Mandelbrot Set

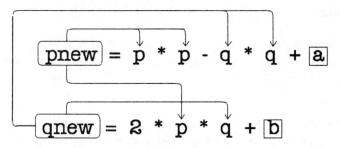

Figure 4.2: Feedback in the Mandelbrot equations

So far we have only specified that a is the initial horizontal position of the point and b is the initial vertical position. No mention has been made of p, p_{new}, q or q_{new}. Naturally these four variables will be subject to change as the equation is applied, and the values of p_{new} and q_{new} will automatically be worked out because they are the results of the equation. That still leaves p and q unaccounted for, and although these will inherit values from p_{new} and q_{new} during subsequent applications of the formula they require some initial values to get the process started. For the time being we shall say that p and q are both set to 0 before the process is invoked, so that the first position calculated is the initial position, (a,b).

Now that we have our set of initial values, and the full equation, we can perform a 'dry run' of the process for one point, as shown below. In this example our starting point will be (-1,0.5), i.e. a=-1, b=0.5.

Initially: $a = -1$, b=0.5, p=0, q=0

Application of the equation gives:

| p_{new} | = $p*p - q*q + a$
= 0*0 - 0*0 + (-1)
= -1 | q_{new} | = $2*p*q + b$
= 2*0*0 + 0.5
= 0.5 |

So the first position is (a,b) as expected.

p now becomes the latest value of p_{new}, and q becomes q_{new}, this can be represented as:

| p | = p_{new} (= -1) | q | = q_{new} (= 0.5) |

The equation can now be re-applied...

| p_{new} | = $p*p - q*q + a$
= (-1)*(-1) - 0.5*0.5 + (-1)
= -0.25 | q_{new} | = $2*p*q + b$
= 2*(-1)*0.5 + 0.5
= -0.5 |
| p | = p_{new} | q | = q_{new} |

...and re-applied...

p_{new}	$= p*p - q*q + a$	q_{new}	$= 2*p*q + b$
	$= (-0.25)*(0.25) - (-0.5)*(-0.5)+(-1)$		$= 2*(-0.25)*(-0.5)+0.5$
	$= -1.1875$		$= 0.75$
p	$= p_{new}$	q	$= q_{new}$

Only three iterations have been performed here to save space, but had more iterations been performed the results in Table 4.1 would have been observed.

Table 4.1: *The first six iterations of the Mandelbrot process for the point with initial coordinates (-1,0.5)*

Iteration number	p	q
0	0.0000	0.0000
1	-1.0000	0.5000
2	-0.2500	-0.5000
3	-1.1875	0.7500
4	-0.1523	-1.2813
5	-2.6184	0.8904
etc...		

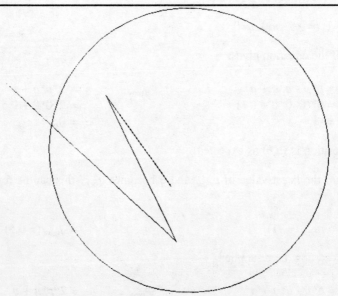

Figure 4.3: *The path of the point with initial position (-1,0.5)*

The Mandelbrot Set

These results are reminiscent of those produced by the virus equation in Chapter 2, in that they are very complex, and appear to be almost random. Of course the results are not random as they come from a deterministic process, which if repeated will yield the same values.

Returning to the wider situation we can now plot the path of the point in the circle, which results in the image shown in Figure 4.3. This output can easily be achieved on the ST by calculating the values of p and q (hence the position of the point) for successive iterations, and then drawing straight lines between these positions. A program to do this is shown in Listing 4.1. Note that the iteration count is displayed in the top left of the screen to give an impression of just how many calculations are taking place.

```
@Check_res
'
Rem Request constant from user
Input "a:",A
Input "b:",B
'
Rem Set up graphics
Cls
Hidem
Circle 160*Monitor,100*Monitor,100*Monitor
'
Rem Set initial values
P=0
Q=0
Iteration=0
'
Rem Execute process
@Fplot(160+A*50,100-B*50) !Plot initial point
Repeat
Pnew=P*P-Q*Q+A
Qnew=2*P*Q+B
P=Pnew
Q=Qnew
@Fdraw(160+Pnew*50,100-Qnew*50) !Draw line to next point
Inc Iteration !Increment iteration count
Print At(1,1);"Iteration:";Iteration
Until Mousek=2 Or P*P+Q*Q>=4 !Check for end of program
@Waitmouse
```

Listing 4.1: Plotting the path of the point in the circle

The program automatically stops when the line breaches the circle, or when the right hand mouse button is pressed. The detection of the mouse event is fairly simple as the built-in GFA BASIC variable `mousek` can be used. Checking whether the point is outside the circle is done mathematically, using the technique shown in Figure 4.4 (overleaf).

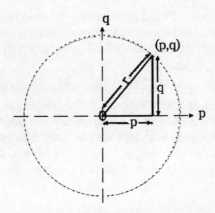

Figure 4.4: Method of testing whether a given point is in the circle

---

The diagram shows that if (p,q) and $(0,0)$ are taken as being two corners of a right angled triangle, the distance between these two corners is equal to the length of the triangle's longest side, the hypotenuse. The length of the hypotenuse (r) can be calculated from the length of the other two sides using Pythagoras' theorem, the formula for which is shown below.

$r^2=p^2+q^2$

If the point (p,q) is inside the circle it is impossible for the hypotenuse to be greater than 2, so:

$r<=2$

squaring both sides gives:

$r^2<=4$

and after substitution into the Pythagoras equation:

$p^2+q^2<=4$

It is this final form of the equation which is used in the UNTIL line to check whether the program should end. Appendix B provides a more detailed description of Pythagoras' theorem.

Because the range of values for p and q is so small the plotting co-ordinates must be significantly amplified in order to make full use of the screen. This is done, as usual, in the `fplot` and `fdraw` lines. Also familiar is the reversal of the q position, required because the ST's screen co-ordinates are numbered from top to bottom rather than in the conventional bottom to top way (see the graphs section of Appendix B for details).

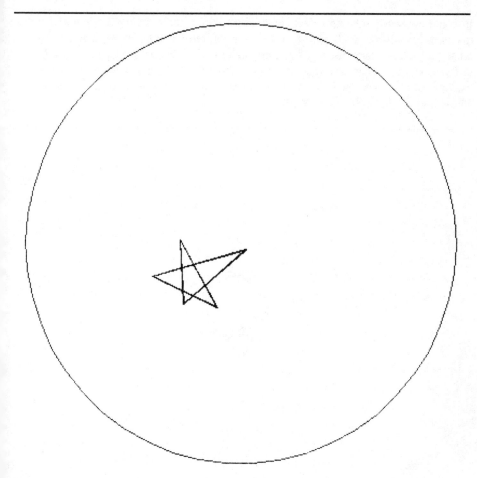

Figure 4.5 (a): Behaviour of the point with initial position (-0.52, -0.55)

From Figure 4.3 it can be seen that the route which the point takes is not especially structured. This is true of most paths plotted by the program, but sometimes patterns do seem to emerge, for example the pentagram in Figure 4.5(a) which is the path of the point with initial position (-0.52,-0.55).

This is all very interesting, and you may want to spend some time experimenting with various initial positions of the point, but how does this relate to the Mandelbrot set? Well, each set of initial values (a and b) can be put into one of two separate categories depending on the path of the corresponding point, generated by applying the Mandelbrot equation. The two categories of points are:

1) Those whose paths rapidly accelerate to infinity

2) Those whose paths never leave the confines of the circle

It would obviously take an infinite amount of time to thoroughly test a point in order to determine whether it flies off to infinity or not, but it can be shown that if the point breaches the boundary of the radius two circle it will eventually reach infinity. Using this reasoning we can say that the initial point (-1,0.5) demonstrated earlier is an example of a category one point. An example of a category two point is (0.35,0.35), whose path is shown in Figure 4.5(b).

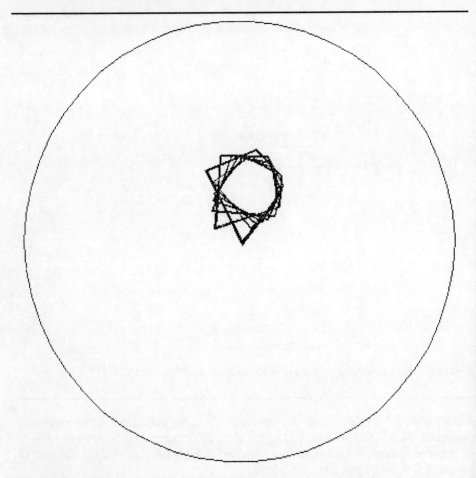

Figure 4.5 (b):Behaviour of the point with initial position (0.35, 0.35)

This was tested using the program above, and had still not exited the circle after 20,000 iterations. Unfortunately there is no simple way to prove that a line will not reach infinity, so the brute force method of applying the equation many times must be used. The point at which we terminate the program and assume the point will never leave the circle is arbitrary, but iteration ceilings of 33 are generally typical (details of how to determine this number are given later).

The Mandelbrot Set

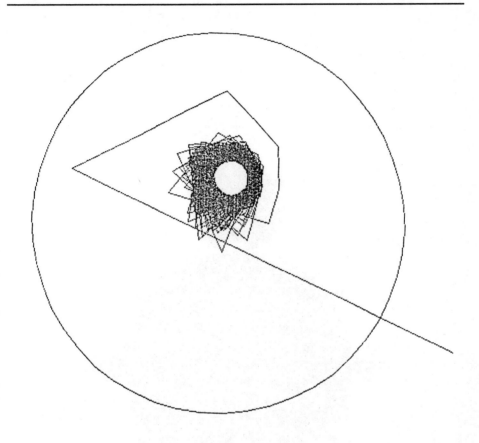

Figure 4.5 (c): Behaviour of the point with initial position (0.3505, 0.35)

To underline the behaviour of the equation Figure 4.5(c) shows another type one line, this time produced from the initial point (0.3505, 0.35). It is interesting to note that just the small change (a is 0.0005 units larger than in Figure 4.5(b)) produces an eminently different result. Such sensitivity to small changes in the initial conditions is typical of non-linear systems, as discussed in the previous chapter.

The categorisation of points according to their behaviour in the circle is the key to creating the Mandelbrot set. What we do is to take every point on another plane, where $-2<=p<=2$ and $-2<=q<=2$, put them through the equations, (as the initial position), and then categorise them. If all the type two points are coloured black and all others are left white we get the image shown in Figure 4.6, commonly known as the Mandelbrot set.

Figure 4.6: Mandelbrot set produced using the 24 pin printing techniques introduced in Appendix D

Figure 4.7 has been included to help visualise the process and shows the relationship between points of the Mandelbrot set and their associated circle planes. A program to produce a plot similar to the one shown in Figure 4.6 is given in Listing 4.2.

The Mandelbrot Set

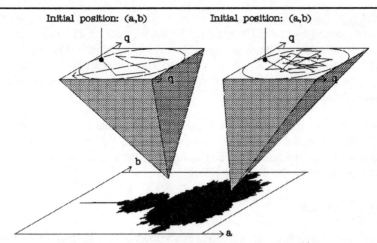

Figure 4.7: Relationship between the Mandelbrot circle and the set

```
@Check_res
'
Hidem
'
For A=-2 To 2 Step 0.02/Monitor
For B=-2 To 2 Step 0.02/Monitor
'
Rem Set initial values of variables
P=0
Q=0
Iteration=0
'
Repeat
Pnew=P*P-Q*Q+A
Qnew=2*P*Q+B
P=Pnew
Q=Qnew
Inc Iteration
Until P*P+Q*Q>=4 Or Iteration=33
If Iteration=33 !If it's a type two point...
@Fplot(160+A*50,100-B*50) !...plot it
Endif
Next B
Next A
@Waitmouse
```

Listing 4.2: A program to produce the plot similar to Figure 4.6

Note that not all the points on the 4x4 plane can be tested because, as on any plane, there are an infinite number, meaning that they would take an infinite amount of time to process and they would not fit on the screen. Instead we plot a small portion of regularly spaced samples, allowing a reasonable approximation of the Mandelbrot set to be built up. An approximation is the best that can be achieved because of the limitations imposed by the ST's screen. The actual number of points calculated is determined by the distance between them. In Listing 4.2 this is the STEP size of the FOR...NEXT loops. This distance is optimised here in order to give the maximum detail in the smallest amount of time, so naturally varies depending on what screen resolution is being used.

In simple terms we are running a simple, two dimensional, strange attractor with a series of different initial conditions. However, the ceiling of 33 iterations for each point means that mathematical feedback has a good chance to work without the problems introduced by the butterfly effect.

Adding Colour

Technically the Mandelbrot set is defined as the map of all the category two points on the plane (i.e. all the points shown in Figure 4.6), but most people are more familiar with the version surrounded by colourful curved lines. These lines are known as contour lines, and are analogous to those found on a map, except that instead of representing height they represent the ease with which the points left the circle. Listing 4.2 can easily be adapted to display these contours. Because the colour of each point represents the ease with which it left the circle, the colour can obviously be deduced from the number of iterations performed before the point escaped - the fewer iterations it took the easier it was.

In the ST's low-resolution mode we can determine the colour by taking MOD 16 of the number of iterations required to free each point from the circle. This gives a range of colours between 0 and 15. Obviously colours are repeated, meaning that no one colour can be assigned to any one iteration number, but the colour cycle ensures that no two adjacent contours are the same. The iteration ceiling of 33 ensures that all members of actual Mandelbrot set are coloured black in line with conventions (33 MOD 16 = 0). For more information on the MOD operator see Appendix B.

To add colour the IF...ENDIF section of Listing 4.2 should be removed, and the following routine substituted in its place:

```
Color Iteration Mod 16  !Set colour
@Fplot(160+A*50,100-B*50)  !Plot point
```

If, however, your Mandelbrot is destined for a monochrome monitor or a printer you will only have a palette of two colours to choose from. In this case the most effective method of displaying the set is to plot a point in black if the number of iterations needed to remove that point from the circle is odd, but otherwise to leave it white. An alternative IF...ENDIF replacement is shown below. Because the maximum iteration number is odd (33), the actual set is automatically coloured black.

```
If Odd(Iteration)
@Fplot(160+A*50,100-B*50)
Endif
```

Figure 4.8 (a):Nine pin low-resolution screen dump of Mandelbrot set with colour contours

Standard nine pin dot-matrix screen dumps showing the output of both versions of Listing 4.2 are given in Figures 4.8(a) and 4.8(b). Figure 4.8(b) most clearly shows the boundaries of the contour lines, and as this book is set in only two colours this monochrome convention will be used in all subsequent Mandelbrot related programs and figures. Naturally Figure 4.8(a) would have benefited from being dumped to a colour printer, and colour monitor users will probably want to replace the monochrome plotting method used in the examples with the 16-colour equivalent described above.

The majority of ST Mandelbrot programs take the best part of 30 minutes to produce the whole contoured set, although exact times depend on the nature of the program and the screen resolution. Hopefully you can now appreciate why they take so long, remember that for each point on the Mandelbrot set the program must determine and categorise the behaviour of the corresponding line in the circle. If we assume that on average 16 iterations are required to categorise each point, and that the size of the relevant screen area is 200x200 pixels (low resolution) then altogether 200x200x16 = 640,000 iterations must take place. That's 640,000 applications of the two equations, the circle breaching test and all the associated program instructions. For a high resolution monochrome plot (400x400 pixels) the number of iterations is around the 2560,000 mark. This is a clear demonstration of why computers are now so relevant to chaos research.

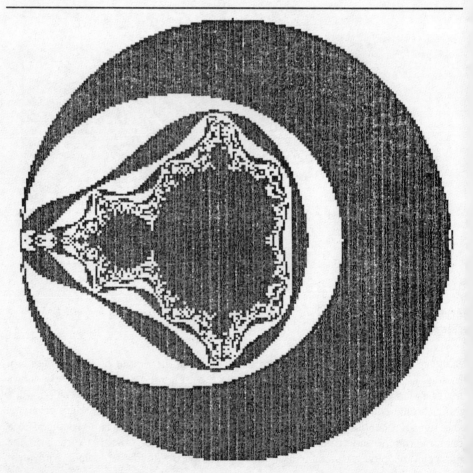

Figure 4.8 (b): Nine pin low-resolution screen dump of Mandelbrot set in monochrome

Of course, not all points require this number of iterations, and if you observe the contour drawing version of the program in action you can easily spot the changes in calculation speed. This is most noticeable in and around the actual Mandelbrot set, where more iterations must be performed per point than anywhere else.

The Complex Number Method

As mentioned in the first part of this chapter one, of the Mandelbrot set's most outstanding features is that such an intricate shape can be created from such simple rules. The circle and line method is not a particularly good way of showing this, as it is rather involved. By contrast the use of complex numbers can allow the Mandelbrot to be created from the single equation:

$$z_{new} = z^2 + c$$

This is a very simple equation with mathematical feedback, much like the virus population one described in Chapter 2. In plain English this equation states that z is equal to the previous value of z multiplied by itself, plus a constant, c. In BASIC the equation might be entered as:

```
z = z*z + c
```

If we repeatedly apply this to normal numbers such as $z=0.5$ and $c=1$ we find that z tends to zero or infinity in a very ordered fashion. However, if complex numbers are used more interesting results can be observed. Before proceeding further a basic understanding of complex numbers may be necessary for some readers.

Complex Numbers

The mere mention of complex numbers can strike fear into many non-mathematicians. The word *complex* in their name is unfortunate because, rather than meaning *complicated*, this actually reflects the fact that such numbers consist of many parts. Complex numbers are nothing more than a mathematical tool, and are not necessarily a key part in the Mandelbrot process, as the circle method shows.

A complex number actually comprises only two parts, called the real and imaginary parts. A typical complex number is shown below, at first it may look like an algebraic equation, but it actually tells us that it is a complex number consisting of four real parts and two imaginary ones:

$4+2i$

Other examples of complex numbers are:

6−4i	6 real and −4 imaginary parts
−2+2i	−2 real parts and 2 imaginary ones
5+i	5 real and 1 imaginary part (i=1i)
0.745+0.113i	0.745 real and 0.113 imaginary parts

The i not only allows us to distinguish the imaginary part from the real part, but also represents the square root of -1 (see Appendix B for details of indices and roots). If you try to calculate the square root of -1 on a pocket calculator using the usual method for square-rooting numbers you will probably be presented with an error message. This is because the roots of negative numbers are impossible to calculate.

Despite this the root of -1 is in fact quite useful in mathematics so mathematicians are not satisfied with the error message on their calculators and instead have assigned the letter i, for imaginary, to it. Several clever things can be done with this number. For example if it is multiplied by itself the answer -1 is obtained, just as multiplying the square root of 4 by itself gives 4. This helps us to manipulate complex numbers, as real numbers can be produced from the imaginary parts.

SQR(4) * SQR(4) = 4

SQR(–1) * SQR(–1) = –1

$i * i = -1$ because $i \equiv$ SQR(–1)

Adding complex numbers is a simple case of gathering together the real and imaginary parts into one number, e.g.

$4+5i + 2+3i = 4+2 + 5i+3i = 6+8i$

$2-i + 3+3i = 2+3 + -i+3i = 5+2i$

Subtraction is similar, except that compatible parts are taken away from each other instead of being gathered together, e.g.

$10+15i - 5+5i = 10-5 + 15i-5i = 5+10i$

$6+3i - 4-2i = 2+5i$ (two minuses give a plus, 3—2 = 3+2 = 5)

Multiplication in the Mandelbrot equation is limited to multiplying a number by itself, often referred to as squaring the number. In general, using the variables a and b to represent the complex number's real and imaginary parts respectively, we can use the equation below to determine the result of squaring a number:

$(a+bi)^2 = a^2 + (2*a*b)i - b^2$

An example of this formula in action is shown below for the calculation of $(4+2i)^2$. $a=4$, $b=2$, therefore the answer is $4^2 + 2*4*2*i - 2^2$

$$= 16 + 16i - 4$$
$$= 12 + 16i$$

Because GFA BASIC, like most other computer languages, cannot deal directly with complex numbers this simple formula will prove invaluable in Mandelbrot-related programs.

Applying Complex Numbers to the Equation

Now that we can add, subtract and square complex numbers we can set about using them in the Mandelbrot equation shown below. The only unknown is which values to put in for c and z.

$z_{new} = z^2 + c$

In the process of plotting the Mandelbrot set, this equation is used to test various complex numbers between $-2-2i$ and $2+2i$, to see if they do one of two things, either they become very large very quickly, or they stay small. A number is tested by placing it into the equation as the constant, c, with z initially set to zero ($0+0i$). An example of a number ($-1+0.5i$) under the first few iterations of the test is shown below.

$$\begin{aligned} z_{new} &= z^2 + c \\ &= 0^2 - 1+0.5i \\ &= -1+0.5i \end{aligned}$$

$z = z_{new}$

$$\begin{aligned} z_{new} &= z^2 + c \\ &= (-1+0.5i)^2 - 1+0.5i \\ &= (-1)^2 + (2*(-1)*0.5)i - (0.5)^2 - 1+0.5i \\ &= 1 - i - 0.25 - 1+0.5i \\ &= -0.25 - 0.5i \end{aligned}$$

$z = z_{new}$

$$\begin{aligned} z_{new} &= z^2 + c \\ &= (-0.25 - 0.5i)^2 - 1+0.5i \\ &= (-0.25)^2 + (2*(-0.25)*(-0.5))i - (-0.5)^2 - 1+0.5i \\ &= 0.0625 + 0.25i - 0.25 - 1+0.5i \\ &= -1.875 + 0.75i \end{aligned}$$

The number is categorised into one of the behavioural groups described in the last paragraph by taking the modulus of the resulting complex number after each stage of the calculation.

The modulus of a complex number is a simple way of describing its size using normal numbers, if the number is represented as $a+bi$, the modulus is the square root of (a^2+b^2). It can be shown that if the modulus of the complex number, z, becomes greater than two then it is destined to become larger very quickly. If the number's modulus stays below 2 for an arbitrary number of iterations, say 33, it is assumed that the number is destined to stay small.

This method of categorisation is applied to a regularly spaced selection of complex numbers between $-2-2i$ and $2+2i$, and all or these which have been assumed to be of the 'staying small' type are plotted on a plane of real against imaginary parts, such as that shown in Figure 4.9.

Note that because BASIC cannot manipulate complex numbers directly they have to be explicitly represented as two separate variable parts as follows:

$z = p + qi$

$z_{new} = p_{new} + q_{new}i$

$c = a + bi$

Incorporating this, and the rest of the process, into a program gives Listing 4.3, which when executed produces the Mandelbrot set.

```
@Check_res
'
Hidem
'
For A=-2 To 2 Step 0.02/Monitor
For B=-2 To 2 Step 0.02/Monitor
'
Rem Set up initial values of variables
P=0
Q=0
Iteration=0
'
Repeat
Pnew=P*P-Q*Q+A
Qnew=2*P*Q+B
P=Pnew
Q=Qnew
Inc Iteration
Until P*P+Q*Q>=4 Or Iteration=33
If Iteration=33
@Fplot(160+A*50,100-B*50)
Endif
Next B
Next A
@Waitmouse
```

Listing 4.3: Program to produce the Mandelbrot set using complex numbers

This listing is exactly the same as the one derived from the circle method (Listing 4.2), proving that both methods lead to the same result. Even the UNTIL line is the same, because the equation for the distance of a point in the circle from the origin is the same as the one for the modulus of a complex number. There is not even a speed difference between the programs. However modern C compilers have extensive

The Mandelbrot Set

complex number handling routines as standard, making this alternative more attractive from a performance and clarity viewpoint.

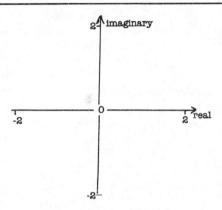

Figure 4.9: The 4x4 complex plane

Manipulating the Mandelbrot Set

The Mandelbrot set is an incomprehensibly complex object, so a single plot cannot possibly contain enough detail to show all the intricacies of the set. For this reason it is useful to be able to mathematically manipulate the set using the ST, something which can easily be done in GFA BASIC.

Zooming In

Most programs written to display the Mandelbrot set allow the user to zoom in and pan around the set, so as to examine certain parts in more detail. This is easy to do using either method, but from now on, for convenience, the process will be discussed in the circle context only.

Assume that we wanted to zoom in on the section shown in Figure 4.10 (overleaf). By intelligent guess-work, or by measuring the diagram, it can be shown that the area in question lies in the region of the plane where $0.25<=a<=0.5$ and $0.5<=b<=0.75$. This is all the information we need to magnify this section, and only three minor alterations need to be made to Listing 4.2 in order to do it. The first is to make the program only test points in the selected region, rather than in the whole plane, in order to do this the two FOR...NEXT loop initialisation lines should be altered to read as follows:

```
For A=0.25 To 0.5 Step 0.02/Monitor
For B=0.5 To 0.75 Step 0.02/Monitor
```

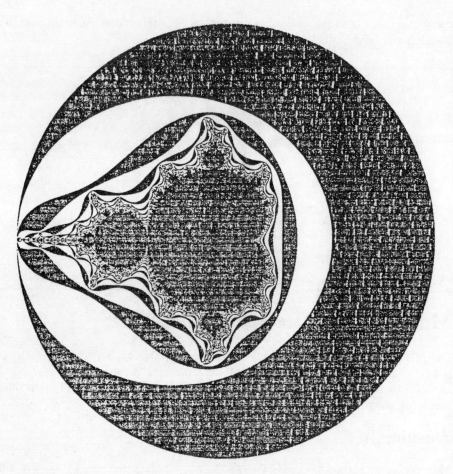

Figure 4.10: 24 pin printout of the contoured Mandelbrot set

This has the desired effect of only showing the relevant area, but the points are plotted in the same place as before, rather than filling up the whole screen. The movement of the points is done by editing the line which draws the points. The relevant replacement for the `fplot` line is:

```
@Fplot(A*800-140,200-(B*800-400))
```

Running the program now will display the right section at the relevant size, but the pixels are very spaced out. This is because the STEPs of the FOR...NEXT loops were optimised for the full, four by four, Mandelbrot set. Because the height and width of our section is only one sixteenth the size of the total Mandelbrot the STEP sizes must be reduced to this proportion. The revised FOR...NEXT initialisation lines are:

```
For A=0.25 To 0.5 Step 0.00125/Monitor
For B=0.5 To 0.75 Step 0.00125/Monitor
```

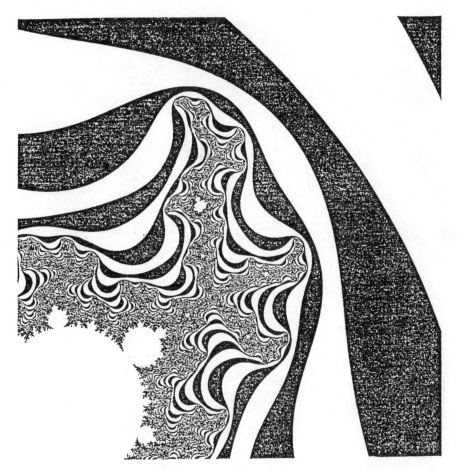

Figure 4.11: Mathematical enlargement of section from figure 4.10 (24 pin). Amin=0.25, amax=0.5, bmin=0.5, bmax=0.75

Figure 4.11 shows the kind of output that you can expect from the altered program. A further enlargement is shown in Figure 4.12. This can be created using the general purpose Mandelbrot program, given as Listing 4.4, in which the values of the `amin`, `amax`, `bmin` and `bmax` variables may be altered in order to display any part of the set.

The section shown in Figure 4.12 (overleaf) represents a 100x magnification of Figure 4.10. If any higher magnification factors are used the maximum number of iterations carried out before assuming the number will never leave the circle (`max_iteration` in Listing 4.4) will have to be increased from 33. This is because, as more magnification is performed, more contour lines become visible, so more calculations must be done to resolve points on the contours from points which actually belong to the 'official' Mandelbrot set.

Figure 4.12: Mathematical enlargement of section from figure 4.11 (24 pin).A Amin=0.34, amax=0.38, bmin=0.63, bmax=0.67

There is no simple formula for calculating the maximum number of iterations for a given part of the set, because the contours are of varying widths. As a rule of thumb the maximum iteration number is good if the border of the actual set is jagged and free of smooth lines. Naturally the lower this number is the better, as the set will take less time to generate.

There is one other important point to make about magnifying parts of any fractal, which is that the aspect ratio (the ratio of height to width) should always be preserved. This is the reason why the sets plotted by Listings 4.2 to 4.4 only take up the middle section of the screen. Extending it to fill the rest of the space would cause it to be artificially elongated. Such tampering with the aspect ratio causes inaccurate plots to be produced, making self-similarity hard to distinguish.

The Mandelbrot Set

Self–similarity is much in evidence in Figure 4.11, which is reassuring proof that the Mandelbrot set is a fractal, in case you didn't already know! The tiny replicas of the Mandelbrot shape are thought to be found even when the set is infinitely magnified, at a lecture Professor Mandelbrot exhibited a picture showing a section of the set magnified by a factor of 10^23 (100,000,000,000,000,000,000,000) times its original size), and still the original Mandelbrot image could be seen.

Most Mandelbrot exploration programs allow the user to select the area to magnify using the mouse, and indeed a program for this purpose is included on this book's support disk. To save repetition, the techniques used in such a program are not explained here. However a fully documented example for similar exploration of Julia sets is included in the next chapter.

```
@Check_res
'
Hidem
'
Rem User editable constants
Amin=0.34 !Lowest value of a
Amax=0.38 !Highest value of a
Bmin=0.63 !Lowest value of b
Bmax=0.67 !Highest value of b
Max_iteration=33 !Iteration ceiling
'
Rem Calculate ranges, offsets and multipliers
A_range=Amax-Amin
B_range=Bmax-Bmin
A_mult=200/A_range
B_mult=200/B_range
A_offset=(0-Amin)*A_mult+60
B_offset=(0-Bmin)*B_mult
'
Rem Draw set
For A=Amin To Amax Step (A_range/200)/Monitor
For B=Bmin To Bmax Step (B_range/200)/Monitor
P=0
Q=0
Iteration=0
Repeat
Pnew=P*P-Q*Q+A
Qnew=2*P*Q+B
P=Pnew
Q=Qnew
Inc Iteration
Until P*P+Q*Q>=4 Or Iteration=Max_iteration
If Odd(Iteration)
@Fplot(A*A_mult+A_offset,200-(B*B_mult+B_offset))
Endif
Next B
Next A
@Waitmouse
```

Listing 4.4: General purpose Mandelbrot program

In his book, *The Fractal Geometry of Nature*, (see bibliography for details), Mandelbrot discusses the likeness between the boundary of the Mandelbrot set and the coastline (land/sea boundary) of an island. Comparing the set (Figure 4.6) with a map of Great Britain does reveal that although the Mandelbrot is symmetrical and contains near-perfect circles, there are similarities. For instance both have thin cracks running into them and thin ligaments coming out, known to geographers as rivers and peninsulas respectively. By far the most stunning feature, however, is the similarity between the jagged nature of the Mandelbrot boundary and the coastline.

We have already seen that after multiple magnifications the Mandelbrot set maintains the same complexity, and theoretically the same is true of the British coastline. If you were to parachute to the ground from a considerable height the amount of 'crinkliness' visible along the coastline would be constant during the whole descent. This is true because more detail is visible as the coastline gets nearer, but at the same time the field of view is reduced. Even when the parachutist lands the level of visible crinkliness will be the same (if he lands near enough to the coastline to see it) because each tiny piece of rock has an outline of equal detail (in relation to its size) to that of the British coastline. Because of this feature, coastlines are classed as fractals, and researchers have even determined a fractal dimension between 1 and 2 for the British coastline. This dimension indicates that the coastline is something between a line, (1 dimensional), and a plane (2 dimensional).

Other Ways of Displaying the Mandelbrot Set

So far the Mandelbrot set has only been plotted on a two dimensional plane, with contour lines showing the ease with which points left the circle. However, this is not necessarily the best way to observe the set, and other methods are available. The first of these is to plot the contours in a more map-like style, using thin monochrome lines rather than wide bands of colour.

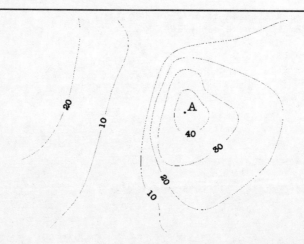

Figure 4.13: Geographical map showing variation in land height (relief)

The Mandelbrot Set

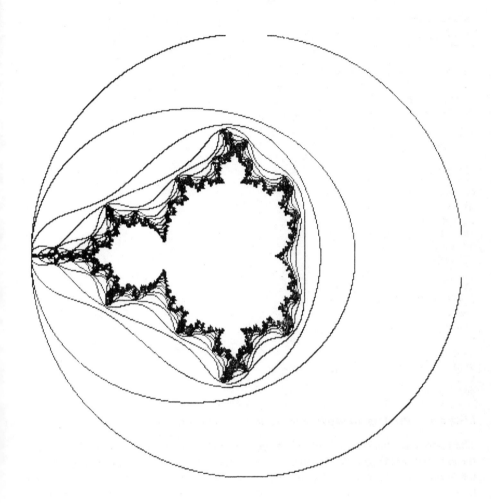

Figure 4.14: Anaemic Mandelbrot set

A typical geographical map section is shown in Figure 4.13, here the lines are used to represent height, with every point on a particular line being at the same height. This allows the height of any particular area to be found, for example it can easily be seen that the point A is at a height of over 40 metres above sea level.

A similar diagram of the Mandelbrot set is shown in Figure 4.14, in this case the lines represent the ease with which the points on the plane left the circle. An 'anaemic' Mandelbrot set of this type can be created using the program given in Listing 4.5. This is basically a re-working of Listing 4.2 which, instead of colouring groups of points depending on the ease at which they left the circle, plots lines to show the boundaries between these groups.

```
@Check_res
'
Hidem
'
Dim Previous_row(200*Monitor)
'
For a=-2 To 2 Step 0.02/Monitor
For b=-2 To 2 Step 0.02/Monitor
p=0
q=0
Iteration=0
Repeat
pnew=p*p-q*q+a
qnew=2*p*q+b
q=qnew
p=pnew
Inc Iteration
Until p*p+q*q>=4 Or Iteration=24
If a>-2 And b>-2
If Iteration<>Previous_row((b+2)*50*Monitor) Or
Iteration<>Previous_row((b+2)*50*Monitor-1)
@Fplot(160+a*50,100-b*50)
Endif
Endif
Previous_row((b+2)*50*Monitor)=Iteration
Last_iteration=Iteration
Next b
Next a
@Waitmouse
```

Listing 4.5: Program to produce the anaemic Mandelbrot set

The boundaries between these groups are detected using a fairly simple algorithm. In the simplest terms a point is plotted if the number of iterations required to free it from the circle is different to that of the points adjacent to it. Because the program traverses the screen from left to right, bottom to top, any point being plotted (providing that it is not right at the edge of the plane) will have the point adjacent to it on the left and the point below it already calculated. To save time these two pre-calculated points are the only adjacent points tested against the point being investigated. However, the iteration numbers associated with these adjacent points were not stored in our original program, because such information was not needed, and it would have required an array with a 160,000 integer capacity (40,000 in low resolution).

In Listing 4.5 a 400 element (or 200 element in low resolution) array is set up at the start of the program to store every value in a particular column. This array is updated whenever a new point is calculated, just before the end of the FOR...NEXT loop. This ensures that for any point the iteration numbers of the two adjacent points are stored in the array, and that as soon as an array element is no longer needed it is overwritten with the latest calculation. This may seem rather complicated at first but a look at Listing 4.5, together with its action, should rectify the situation.

Enhancements to the Mandelbrot Programs

Because it is difficult with only 16 colours (and impossible with only two) to give each contour around the Mandelbrot set a unique colour it would be useful to have another way of representing these contours. A popular method for doing this is to draw an isometric three dimensional Mandelbrot, known as a Mandelbrot landscape, where points are elevated from the plane by differing amounts depending on the ease with which they left the circle. Because isometric drawing is an essential element of this technique 3D Mandelbrots are discussed in the chapter on fractal landscapes (Chapter 7).

Internal Structure

In the figures shown in this chapter the area inside the Mandelbrot boundary has always been coloured black, but it is actually possible to colour the inside of the set with contours similar to those found outside. Although these internal contours are derived from the Mandelbrot process they have little importance to the study of the set, and are simply included to add visual effect. I do not intend to discuss internal structure in detail, but Figure 4.15 has been included to give an in idea of how it manifests itself in the Mandelbrot set.

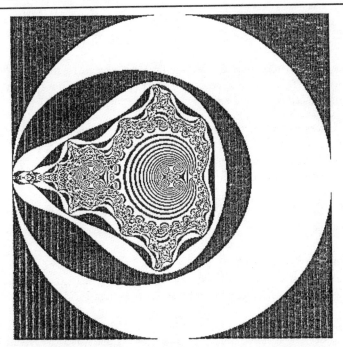

Figure 4.15: *Mandelbrot set with internal structure*

5

Julia Sets

By looking at the Julia set plots included in this chapter it is clear that there are many similarities between these and the more familiar Mandelbrot set. This is because the same equation is employed to draw both fractals. It is merely the way in which it is applied that make them different. This close link between the two fractals means that Julia sets provide just as much scope for experimentation as their more famous counterpart.

The Julia Process

Both methods used for drawing the Mandelbrot set, shown in the last chapter, are suitable for drawing Julia sets. However, the circle method will be used here in order to avoid the intricacies of complex numbers.

In Chapter 3 it was suggested that the Lorenz attractor was based on three equations containing three variables, x, y and z, so the system could be treated as being three dimensional. In Chapter 4 we found that to plot the path of the point (p,q) with initial position (a,b) in the Mandelbrot circle the following pair of equations was required:

$p_{new} = p^\wedge 2 - q^\wedge 2 + a$

$q_{new} = 2*p*q + b$

This system clearly has four variables (p_{new} and q_{new} are not counted as they become p and q on subsequent iterations). Following the theory of the Lorenz equations it is clear that the Mandelbrot set can be treated as a four dimensional object.

While it is challenging enough to represent three dimensional objects on a flat screen, displaying a four dimensional object is practically impossible. To overcome this problem only two variables from the Mandelbrot equation are used to plot the set.

Julia Sets

These are *a* and *b* which are used to position each point – *p* and *q* are used in the calculations but are later discarded.

Figure 5.1 shows the relationship between the line in the circle for a Mandelbrot set, and next to it the relationship between a similar line in a circle and a Julia set. In the case of the Julia set the initial position of the point in the circle (*a,b*) is kept constant for the whole set, and the variable initial values of *p* and *q* are used to determine the position of the point to be plotted. Apart from this subtle change the rest of the process is the same for both fractals, including the test for escape from the circle, the method of determining the colour of contours and the method used to magnify a section of the set.

You may have noticed that Julia sets keep being mentioned in plural form. This is because there are an infinite variety of possible sets which can be produced using the Julia process, each with a unique initial position constant (*a,b*). In most chaos literature this constant is expressed in the equivalent complex number form of *a+bi*.

The example program in Listing 5.1 draws the Julia set with constants *a*=-1.16 and *b*=-0.25 (-1.16-0.25i).

```
@Check_res
Hidem
'
Rem Set constants
A=-1.16
B=-0.25
'
Rem Plot set
For X=-2 To 2 Step 0.02/Monitor
For Y=-2 To 2 Step 0.02/Monitor
P=X
Q=Y
Iteration=0
Repeat
Pnew=P*P-Q*Q+A
Qnew=2*P*Q+B
P=Pnew
Q=Qnew
Inc Iteration
Until P*P+Q*Q>=4 Or Iteration=33
If Iteration=33
@Fplot(160+X*50,100-Y*50)
Endif
Next Y
Next X
@Waitmouse
```

Listing 5.1: Program to draw the Julia set with the constant –1.16-0.25;

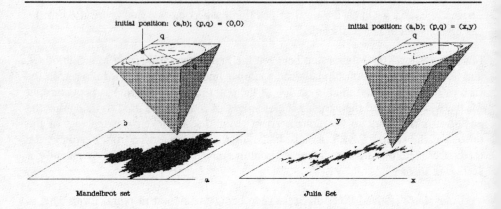

Figure 5.1: Comparison of Julia and Mandelbrot processes

In many respects this program is similar to the equivalent Mandelbrot example given in Listing 4.2. Because the Julia plane extends from −2 to 2 in both directions, as in the Mandelbrot set, even the scaling is the same. The main differences are that a and b are set as constants at the start of the program and p and q, instead of being set to zero at the start of the circle process, are given values (x and y) derived from the nested FOR...NEXT loops. This program can be made to draw any Julia set simply by changing the values of the constants a and b, some interesting values are suggested in Table 5.1.

Table 5.1: Some suggested Julia set constants

a	b	Equivalent complex number
-1.16	-0.25	-1.16-0.25i
0.32	0.04	0.32+0.04i
-1.25	-0.01	-1.25-0.01i
0.00	-1.00	-i

Adding Contours

Contours can be added to Julia sets using the same methods used in the previous chapter for the Mandelbrot set, where the contours are created by colouring points depending on the ease with which they left the circle. To save you referring back to Chapter 4, the possible replacements for the IF...ENDIF section of the program are as follows:

For 16 colours (low resolution only):

```
Color Iteration Mod 16  !Set colour
@Fplot(160+X*50,100-Y*50) !Plot point
```

For monochrome contours (high or low resolution):

```
If Odd(Iteration)
@Fplot(160+X*50,100-Y*50)
Endif
```

Like the Mandelbrot set, Julia sets can also be drawn as 3D landscapes (see Figure 5.4), or with internal structure.

Derivation from the Complex Number method

Again, the method is similar to that given for the Mandelbrot set. The equation shown below still holds but c is kept constant the whole time, rather than being different for each point on the complex plane.

$$z_{new} = z^2 + c$$

Because c is constant the position of the point is determined by the initial value of z which, instead of always being zero (0+0i), is different for each point on the plane. The Julia set drawn is dependant on the complex constant, c, whose real and imaginary parts are equivalent to the circle method's a and b variables respectively.

A Julia Set Explorer Complete with Bells and Whistles

In general I have attempted to write the programs for this book in as compact a form as possible. However, there are programs available, many of which are on this book's support disk, which will allow you to explore fractals with all the convenience offered by graphical user interfaces, disk filing operations and printout facilities. This type of program tends to be very long and hence is unsuitable for a book primarily written as an introduction to the theory of chaos. However, as an insight into how such programs are created, a GEM-based example to generate and explore Julia sets is presented in this chapter. This will prove especially useful if you want to develop your own fractal exploration programs, based on processes given in this book or elsewhere.

The full Julia program is shown in Listing 5.2. This is provided mainly for educational purposes and to allow important techniques to be gleaned. Due to the length of this listing it would be irresponsible for me to recommend typing it in. If you are interested in seeing the program in operation I suggest you purchase the support disk instead as this contains the entire program, and a much faster compiled version of it.

Planning

Before actually tackling a project of this sort a clear set of objectives and a plan for developing the program must be devised. The objectives for the Julia program are that it should:

❏ Be able to plot (quickly) any Julia set requested by the user

❏ Allow zooming (of any magnification) of any section of the set

❏ Give options for filing sets and sections to and from disk

❏ Fully utilise special features of peripherals (e.g. monitors)

❏ Display parameters in a form compatible with other literature

❏ Allow for the filing of such parameters for later use

❏ Give the user the option to dump images to a printer

❏ Be implemented in an intuitive graphical environment.

This list basically brings together the Julia set facilities discussed above and the features that would be expected from a commercial art package or word processor. One of the most important points here is the parameter saving option as this would allow (for example) a Julia landscape program to draw three dimensional versions of sets already plotted and saved in two dimensions.

The intuitive graphical environment mentioned in the list must be GEM, as it is resident in the ST ROM and is supported by GFA BASIC. The use of GEM in the Julia program is limited to a pop down menu bar, alert boxes, simple windowing and mouse input. This is because, using version two of GFA BASIC, more sophisticated activities can only be undertaken using relatively complicated techniques. The more expensive GFA BASIC 3.5 (not covered in this book) provides enhanced GEM support of a similar standard to that found in good C systems, so would be slightly more suited to this task.

A discussion of the menu bar (given later) provides a more detailed breakdown of the options available because, in a program of this sort, all the options must be selected either directly or indirectly from the menu bar.

Universal Programming Techniques

There is no escaping the fact that Julia set generators must perform many time-consuming calculations. For this reason a number of steps have been taken to ensure that the Julia program is as fast as possible. The most general is the application of structured programming techniques, including the use of local variables wherever possible. Unlike global variables, these are local to a particular procedure and are lost when that procedure ends, which means that at any time there are fewer variables for BASIC to keep track of than there would be if all the variables were global. Also, when dealing with whole numbers, integer variables have been employed instead of the floating point ones selected by default. Integer variables are signified by having their names suffixed by a per cent character (%) and are processed significantly faster than their floating point counterparts.

I have used comments and long procedure and variable names throughout the program to make it easier to understand but, when running under the interpreter, these will slow the program's execution. Obviously the comments may be omitted, but shortening names can introduce errors if sufficient care is not taken. If you intend to compile the program the long names and comments will not matter because these are automatically stripped out when creating the .PRG file.

The Menu Bar

Because each menu option is consecutively numbered it is imperative that all possible options are considered before incorporating the menu into a program. The conventional way of organising GEM menu bars is for the *Desk* menu to come first, followed by the *File* menu and then a series of program-specific menus. The *Desk* menu should contain an *About* option to display information about the program as well as provide sufficient space for six desk accessories. The *File* menu always contains "File manipulation" options, some of which may be program-specific, followed, after a blank line, by an option to "Quit the program". These conventions have been adhered to in the Julia program, for which the full menu structure is shown in Figure 5.5. The first program-specific menu is the *Picture* menu, whose options are fairly self-explanatory. The *Colours* menu is used to select which of the three methods will be used to display the set, so the three options are mutually exclusive. The action of each option is:

Mono: The set is drawn with mono contours (default)
Set Only: Only members of the actual set are drawn, in black
Colour: The set is drawn with colour contours.

A tick is placed next to the menu item representing the selected colouring method. To save accidental loss of pictures, a tick is also used in the *File* menu (next to the *Save* option) to inform the user of whether or not the set in memory has been saved.

GFA BASIC stipulates that the menu structure must be stored in an array with each menu item in a separate element of the array, placed consecutively in the order in which they are to appear. An empty string ("") is used to mark the end of each menu, and a pair of empty strings mark the end of the last menu (the end of the menu structure). The information for the Julia menus is placed in an array called menu$ in the initialise procedure (see Listing 5.2). This array could have been filled using more compact code, but by using the longhand method it is easy to determine how the menus are arranged and what number each menu item has, merely by glancing at the program.

A further array, called status%, is used to store information about the status of menu items – whether they are disabled for example. This information must be stored in this way because the built-in GFA BASIC menu routines lose such information when the menu bar is switched off (as it frequently is in this program). The status of a menu item is set using a command of the form:

Status%(item_number)=flag

where item_number is the number of the menu item and flag is one of the following values:

0	Enabled, no tick
1	Ticked
2	Light (disabled)

The default status for some items is set just after the initialisation of the menu structure. Of course the contents of the array are not passed to the GFA BASIC menu handler automatically. This job is done by the menu_status procedure which uses a version of the MENU command to set the menu item status. This procedure must be called whenever the menu is switched on after being off, or whenever an item's status is altered while the menu is turned on. Note that menu_status checks the screen resolution and, if necessary, automatically disables the *Colour* option from the *Colours* menu.

Menu Handling

The menu bar is set into action near the end of the initialise procedure. Here the menu is turned on, given the latest status information using a call to menu_status, and instructed to call the branch procedure upon a menu item being selected. The screen information memory blocks used by the file options (see Appendix B for details) are also initialised at this point in the procedure.

After the initialisation procedure is complete an endless DO...LOOP loop is entered which continually checks for menu activity using the ON MENU command. On selection of an item (i.e. when the user has pulled a menu down and clicked on one of the options) the branch procedure is automatically called by the GFA BASIC menu handler.

Julia Sets

When called the `branch` procedure immediately switches off the menu bar to stop confusion occurring due to options being selected while other options are being carried out. The procedure then branches off to one of 11 different procedures depending on the item selected (the item's number is returned by the GFA BASIC menu handler in element zero of the built-in `menu()` array).

On returning from the selected procedure the menu is switched back on and its status updated by calling the `menu_status` procedure. The `branch` procedure then returns control to the menu testing loop at the head of the program. The various procedures called from `branch` perform all the program's main operations and are discussed below.

Desk_About

This very simple procedure is called to display an alert box proclaiming the name and source of the program whenever the *About this program* option is selected from the desk menu. The procedure first initialises the local integer variable, `button%`, before calling `bar_text` to erase the remnants of the menu bar and display a description of the action selected instead. After this and a quick piece of house-keeping the procedure's main function, the creation of the alert box, is carried out. This is performed using the GFA BASIC ALERT command for which the syntax is:

```
Alert Icon,Message$,Default,Button$,Button%
```

where:

`Icon` determines the icon drawn in the box. It can be set to zero for nothing, 1 for an exclamation mark (!), 2 for a question mark (?) or 3 for an octagonal stop sign.

`Message$` holds the message to be displayed, in which the vertical bar character (|) is used to mark the end of each line (and the beginning of the next).

`Default` holds the number of the default button, which will be displayed with a thickened outline and automatically selected if the Return key is pressed while the alert box is on the screen. It is possible to have no default button by placing a zero here.

`Button$` holds the text for one, two or three buttons, separated by vertical bars (|).

`Button%` is a variable in which the number of the chosen button is returned after it has been selected. Note that the buttons are numbered consecutively as they appear in `button$`, starting from 1.

The ALERT command has been explained in detail because it is widely used in the Julia set program for conveying messages to the user. It is particularly useful in a program of this sort because the screen area obscured by the alert box is automatically restored when the box is removed.

The structure of this procedure is similar to all procedures called as a result of a menu item being selected, with the declaration of local variables and the calling of `bar_text` prior to performing the selected actions.

Dummy

Calls to `dummy` are included in the ON GOSUB line of the `branch` procedure to fill in the gaps where no procedure needs to be called (e.g. where unselectable dotted lines are included as menu options). As careless typing could cause `dummy` to be called instead of the correct procedure, a procedure definition for `dummy` has been included which displays an alert box warning of the error.

File Operations

Creating, or zooming into, a Julia set takes a relatively long time and it is therefore useful to be able to save the image on the screen to disk. This is very easy to do using the `degas_save` and `degas_load` procedures given in Appendix A, but the information about the picture, the constant used to create it for example, is only stored in RAM. Therefore it is impossible to identify the picture when it is re-loaded at a later date. In addition, the program does not then have sufficient data to allow the user to manipulate the set.

Obviously this problem could be solved by saving a second file containing the relevant constants whenever a picture is saved, but this would create a rather untidy situation where each plot has two files associated with it. Alternatively, the Julia program could store the pictures and constants in the same file. The fundamental idea

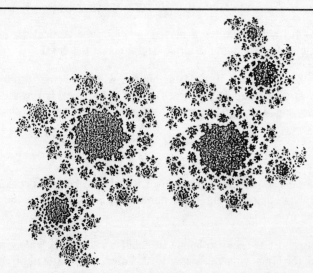

Figure 5.2: Julia set produced using the 24 pin printing methods discussed in Appendix D

Julia Sets

Figure 5.3: Julia set with a=0.32, b=0.043 (24 pin)

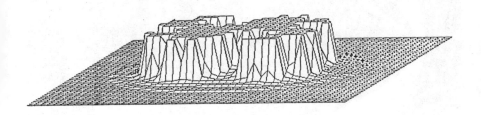

Figure 5.4: Julia landscape produced using techniques from Chapter 7

Desk	File	Picture	Colours
About this program	Load	Plot Julia set...	√ Mono
	Save	Zoom in	Set only
	Delete	View parameters	Colour
	Quit	Print	

Figure 5.5: The menu structure of the Julia set exploration program

behind this technique is to save the picture in Degas format as before, but to include a string of constants at the end of the file. Because Degas, and other programs which adhere to this format, do not expect information at the end of the file it passes through undetected.

The actual format used for the Julia files is shown in Figure 5.6. Naturally the constants for the set are saved, along with the position on the plane from which the saved section was taken. Other information includes the iteration ceiling, the `increase` plotting constant (described later) and a code denoting the file type. The code is not a secret device to stop piracy or anything of that nature, but is simply used by the program to ensure that the file is a Julia set before it is loaded onto the screen. This must be done to stop users trying to load incompatible files, such as standard Degas pictures, which look very similar in the file selector due to their identical PI extensions.

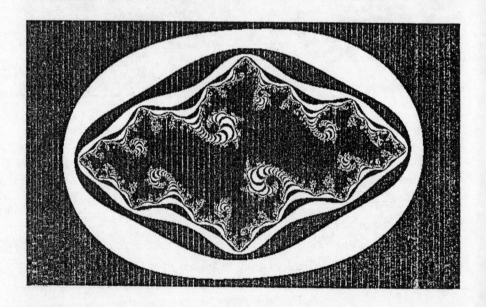

Figure 5.6: File format used in the Julia set exploration program

The code for a Julia file is 808. The other fractal generators supplied on the support disk save files in a similar format with their own unique codes. This information on file storage is essential to the discussions on file handling procedures that follow.

File_load

If the status of menu option 12 is set to zero (if the Julia set currently in memory has not been saved) confirmation of the menu selection is sought using an alert box. This step is necessary as loading a new file overwrites the set previously in memory. If the user confirms the intention to load a new file a file selector is displayed using the FILESELECT command, the syntax for which is:

`FILESELECT searchpath$,suggestion$,selected$`

where:

`searchpath$` contains the path to the directory (folder) in which the file is to be found and the type of file to load. A typical searchpath might be `"C:\CHAOS\PIC-TURES\*.PI3"` which would cause the file selector to display all files with the PI3 extension in the PICTURES directory of the CHAOS directory on drive C.

`suggestion$` contains a suggested filename which the file selector submits for editing. This is usually an empty string when loading files, and a variation of "UNTITLED" when saving.

`selected$` is a variable in which the full name (including the path) of the selected file is returned. An empty string is returned if the cancel button is hit.

After a file has been selected its extension is tested against the contents of `ext$`, generated by the modified `check_res` procedure at the beginning of the program, to ensure the file is of the right type and that it was saved in the resolution being used. If the extension is not correct the file selector is displayed again and the user is given a second chance.

Once a file has been successfully selected the program tests to see whether it exists and, if it does, it goes on to open it for input and checks the code. Providing that the code is correct (it should be 808) the file's constituents are loaded into the relevant parts of memory. After successfully loading the file the *Zoom in* and *View parameter* menu options are enabled, and a tick is placed next to the *Save* option (obviously if the file has been loaded from disk it must already have been saved). If the file does not exist an alert box is generated to inform the user of the problem.

File_save

This procedure is very similar to `file_load` but many of the checks are not performed as we know that the file will exist and can ensure that the correct code is used, because the disk is being written to. However, the file extension returned from the file selector is checked against `ext$` in order to force the user into typing the correct extension for the mode being used. After the file has been successfully saved a tick is placed next to the *Save* menu item (item 12) by setting the 12th element of the `status` array to 1. Note that there is no disk full or write protection check, and any file of the same name as the one being saved will be overwritten without warning.

File_delete

It is sometimes necessary when saving Julia plots to delete old files to make room for the new one. This can leave the user in a bit of a predicament if a *Delete* menu option is not included. The operation of this procedure is very simple, relying on the FILESELECT command discussed above for file selection and using the KILL command to delete the selected file.

File_Quit

This procedure is called when the user indicates the desire to leave the program, by selecting the *Quit* option from the *File* menu. Because of the severity of this action a confirmatory alert box is displayed before releasing the used memory and returning to the GFA BASIC editing screen (returning to the desktop if the program has been compiled).

Picture_julia

Apart from loading a file from disk the only way to get a Julia set on the screen is by selecting the *Plot Julia set* option from the *Picture* menu. `Picture_julia` is the procedure called when this option is selected. If there is an unsaved set on the screen when the option is chosen the program produces a confirmatory alert box before going on to set the `xmin`, `xmax`, `ymin`, `ymax` and `max_iteration%` variables to their default values for a full set plot. A window is then displayed in the bottom left of the screen in which the user must enter the two parts of the constant required to draw the set. The window is opened with the `open_window` procedure and closed after use by a call to `close_window`. The *Save*, *Zoom* and *View parameters* menu items are then enabled, as they now represent viable options, and the `plot_set` routine is called to draw the set. `Plot_set`, `open_window` and `close_window` are described later.

Picture_zoom

In many respects this procedure is similar to `picture_julia` because it initiates the drawing of a Julia set using the `plot_set` routine. The differences are that the constant need not be entered because it is inherited from the set already on the screen, but the dimensions of the set and the iteration ceiling are selected to determine the section of the set instead. The user selects the area to enlarge simply by moving to the relevant area, holding the left mouse button down and dragging an 'elastic' rectangle out from the point until the desired area is enclosed (pressing the right hand button will cancel the process at this stage).

This action is simple from the user's point of view but, behind the scenes, there is a lot of activity. As well as animating the rectangle in such a way that it does not distort the set already on display the program also keeps the aspect ratio of the rectangle the same as that of the screen. Therefore the magnified section of the set does not have to be artificially elongated when plotted. Once the rectangular area has been selected a series of calculations is performed to determine the values of the xmin, xmax, ymin and ymax variables for the new plot and a value for max_iteration% (the iteration ceiling) is determined. The value of max_iteration% is increased proportionally to the level of magnification because the more detailed the plot the more calculations are required to distinguish between different types of points.

The rule used to calculate the iteration ceiling was arrived at empirically because the complex nature of Julia sets prevents us from ever knowing accurately the amount of detail likely to be visible in the enlarged section. Also the result of (max_iteration% MOD 16) must always be zero so that the set is black in the centre, in line with the accepted conventions.

Picture_parameter

As well as displaying the constant for the set currently on the screen (and its position on the Julia plane) this procedure also displays the iteration ceiling, in an editable text field. This latter option is very useful as it allows the user to alter the level of detail resolved, either to speed up plotting or to by-pass the somewhat inaccurate method of calculation used when zooming into a set.

The procedure uses open_window and close_window (both discussed shortly) to manipulate the window in which the information is displayed. This window is shown in operation in Figure 5.7 (overleaf). Two notable GFA BASIC functions are used in picture_parameter: PRINT USING which is employed to format the floating point numbers, and FORM INPUT which places a string on the screen, allows the user to edit it and then returns the result. Note that some conversion between strings and numeric variables (using STR$ and VAL) must be performed here because the variable being edited, max_iteration%, is numeric and FORM INPUT will only accept strings.

Colour

The colour procedure deals with all three options of the *Colours* menu, and is used to select the way in which the set is displayed. The structure of the procedure differs slightly from that of others called from the menu handler, in that no message is placed in the menu bar because the speed of the procedure is such that the user would not have time to read it. After initialising the local variables a FOR...NEXT loop is entered in which the selected option has a tick placed next to it by amending the relevant status% entry. The other two options have their status set to zero (unticked). This loop ensures that only one option can be ticked at a time. Note that this procedure in no way alters the way in which the set is drawn, it merely changes the status array to allow the plotting procedure to know which option should be used the next time it is called.

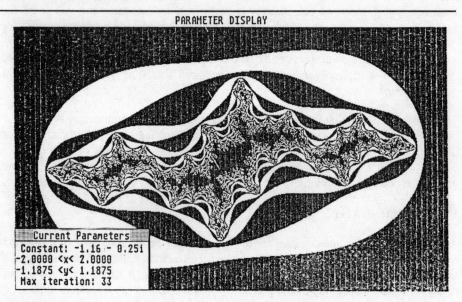

Figure 5.7: *The picture_parameters window in operation*

Procedures not called directly from the Menu Handler

Routines to perform common operations, such as printing messages in the menu bar, have been written as procedures so that they can be easily used in many different program locations. Such routines are described here.

Plot_set

This can be thought of as a universal Julia set plotter as it can draw any section of any set with any iteration ceiling, using any of the three main colouring methods. The procedure fetches the information about the set from the global variables, of which a and b describe the constant, xmin, xmax, ymin and ymax give the position in the set of the section to be drawn and max_iteration% holds the iteration ceiling. The colouring method is determined by interrogating the status% entry for each of the three *Colours* menu options to identify which one is ticked.

Note that an important variable, increase, is calculated before plotting begins. This variable holds the size of a screen pixel in terms of the 4x4 Julia set plane and is used when converting the position of a point on the screen to the corresponding point on a Julia set. Increase is a global variable because, as well as being used in plot_set, it is used in picture_zoom to convert the co-ordinates of the rectangle drawn by the user into the position of the co-ordinates of the Julia set section. As increase is such an important global variable, it is included in all files created by the Julia program.

Check_res

This is the same basic routine as that used in other programs in this book but it returns the resolution multiplier as an integer variable, monitor%, to speed up the plotting of the Julia sets. Also, because pictures are being saved and loaded, this procedure has been changed so that it returns (in ext$) the relevant Degas filename extension for the mode being used.

Round

A function used to round off floating point numbers to the nearest integer. The BASIC INT function cannot be used for this purpose because it is only able to round numbers down. Round is used when converting into integers the floating point numbers produced by the scaling calculations for the PLOT command in plot_set. Note that the fplot and fdraw procedures used in the rest of the book contain their own integrated rounding code.

Open_window and Close_window

Together these provide all the necessary functions for manipulating the parameter/input window. It is relatively easy to open a window in GFA BASIC using the OPENW command, whose syntax is:

Openw window_number,x,y

where window_number holds the number of the quadrant in which the window is going to appear and (x,y) is the position of the point at the centre of the four quadrants, as shown in Figure 5.8. The window created by the open_window procedure is always in the third quadrant.

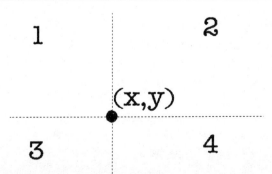

Figure 5.8: Window quadrants in GFA BASIC

The drawback of the standard GFA BASIC window routines is that when a window is closed, using the CLOSEW command, the area previously obscured by the window is not restored to its original state. To overcome this limitation the `open_window` procedure stores the background section of the screen in a global variable, using the GET command, before opening the window. When the window is later closed by calling `close_window`, the screen section is taken from the global variable and placed back on the screen using the PUT command. The syntax for the GET and PUT commands is as follows:

```
GET left_x,top_y,right_x,lower_y,store$
PUT left_x,top_y,store$
```

where `left_x`, `right_x`, `lower_y` and `top_y` describe the rectangular section of the screen in pixels positions and `store$` is the variable used to store the data from that section.

In addition to keeping the background in tact, the `open_window` procedure also gives the window a title taken from the argument that was passed when the procedure was called.

Memory_initialise, palette_store and memory_free

These are the same as the similarly named procedures given in Appendix A. Note, however, that no provision has been made for a printer dump (except for the menu option). However, relevant program examples and advice on printers can be found in Appendices A and D.

```
Gosub Check_res !Check screen resolution and set monitor variable
'
Gosub Initialise
Do
On Menu
Loop
'
Procedure Initialise
Rem The following lines set up the menu options in the A$ array
Dim Menu$(29)
Dim Status%(29)
Menu$(0)=" Desk "
Menu$(1)=" About this program "
Menu$(2)="---------------------"
Menu$(3)="1"
Menu$(4)="2"
Menu$(5)="3"
Menu$(6)="4"
Menu$(7)="5"
Menu$(8)="6"
Menu$(9)=""
Menu$(10)=" File"
```

```
Menu$(11)=" Load"
Menu$(12)=" Save"
Menu$(13)=" Delete "
Menu$(14)="----------"
Menu$(15)=" Quit"
Menu$(16)=""
Menu$(17)=" Picture "
Menu$(18)=" Plot Julia set... "
Menu$(19)=" Zoom in"
Menu$(20)=" View parameters"
Menu$(21)="--------------------"
Menu$(22)=" Print"
Menu$(23)=""
Menu$(24)=" Colours "
Menu$(25)=" Mono"
Menu$(26)=" Set only "
Menu$(27)=" Colour"
Menu$(28)=""
Menu$(29)=""
'
Rem Set up initial menu option status
Status%(12)=2
Status%(19)=2
Status%(20)=2
Status%(22)=2
Status%(25)=1
'
Deffill 1,2,4
Pbox -1,-1,640,400 !Fill screen area
'
Menu Menu$() !Turn the menu on
Gosub Menu_status !Set new status
On Menu Gosub Branch
@Memory_initialise
Return
'
Procedure Branch
Menu Off
On Menu(0) Gosub
Desk_about,Dummy,Dummy,Dummy,Dummy,Dummy,Dummy,Dummy,Dummy,File_
   load,File_save,File_delete,Dummy,File_quit,Dummy,Dummy,Picture_julia,
   Picture_zoom,Picture_parameters,Dummy,Picture_print,Dummy,Dummy,
   Colour,Colour,Colour
Menu Menu$()
Gosub Menu_status
Return
'
Procedure Desk_about
Local Button%
Gosub Bar_text("ABOUT THIS PROGRAM")
Alert 1,"Julia Set Generator|(c)1991 Conrad Bessant",1," OK ",Button%
Return
```

```
Procedure File_load
Local Filename$,Button%,Code%
Gosub Bar_text("LOAD FILE")
'
Button%=1 !Default button is 'Yes'
If Status%(12)=0
Alert 3,"Are you sure you want to|load a new file?|Current one will be
    lost.",2," Yes | No ",Button%
Endif
'
If Button%=1
Repeat
Fileselect "*."+Ext$,"",Filename$
If Right$(Filename$,3)<>Ext$ And Filename$<>""
Alert 1,"That file is not compatible|with this program",1," Retry
    ",Button%
Endif
Until Filename$="" Or Right$(Filename$,3)=Ext$
If Filename$<>""
If Exist(Filename$)
Hidem
Open "I",#1,Filename$ !Open file (read only)
Seek #1,32034
Input #1,Code% !Check file code
Seek #1,0
If Code%<>808 !808 denotes Julia set
Alert 1,"That is not a Julia set",1," Retry ",Button%
Close #1
Else
'
Bget #1,Screen_info,34 !Read resolution and palette
Bget #1,Xbios(2),32000 !Read picture
Input #1,Code%,A,B,Xmin,Ymin,Xmax,Ymax,Max_iteration%, Increase !Load
    constants
Close #1 !Close file
'
For Register=0 To 15 !Load and..
Setcolor Register,Dpeek(palette+2*Register) !..set palette
Next Register
'
Status%(12)=1 !Tick save option
Status%(19)=0
Status%(20)=0
'
Endif
Else
Alert 2,"File not found.",1," OK ",Button%
Endif
Endif
Endif
Return
'
```

```
Procedure File_save
Local Filename$,Button%
Gosub Bar_text("SAVE FILE")
Repeat
Fileselect "*."+Ext$,"Untitled."+Ext$,Filename$
If Right$(Filename$,3)<>Ext$ And Filename$<>""
Alert 1,"The filename must have|the extension ."+Ext$,1," Retry  "
    ,Button%
Endif
Until Filename$="" Or Right$(Filename$,3)=Ext$
'
If Filename$<>"" !If cancel wasn't pressed
Hidem
@palette_store
Open "O",#1,Filename$ !Open file (as output)
Bput #1,Screen_info,34 !Write resolution and palette
Bput #1,Xbios(2),32000 !Write picture
Write #1,808,A,B,Xmin,Ymin,Xmax,Ymax,Max_iteration%,Increase
Close #1 !Close file
Status%(12)=1
Endif
Return
'
Procedure File_delete
Local Filename$,Button%
Gosub Bar_text("DELETE FILE")
Repeat
Fileselect "*.*","",Filename$
If Exist(Filename$) !If it exists...
Kill Filename$ !...delete it
Else
If Filename$<>"" !If not: say so
Alert 3,"File not found",1," OK ",Button%
Endif
Endif
Until Filename$="" !Until 'cancel' is pressed
Return
'
Procedure File_quit
Local Button%
Gosub Bar_text("QUIT PROGRAM")
Alert 3,"Are you sure you want|to leave the program?",2," Yes | No "
    ,Button%
If Button%=1
@Memory_free !Free allocated memory
Edit !Quit back to editor
Endif
Return
'
Procedure Picture_julia
Local Value$,Value%,Button%
Button%=1
If Status%(12)=0
Alert 3,"Are you sure you want to|plot a new set? The current|one will be
    lost!",2," Yes | No ",Button%
Endif
```

```
If Button%=1
Rem The next five lines set the global variables for a full set display
Xmin=-2
Xmax=2
Ymin=-1.1875
Ymax=1.1875
Max_iteration%=33
Gosub Bar_text("ENTER COMPLEX CONSTANT FOR THE SET")
'
Gosub Open_window("Please Enter Constant")
Input "Real part(a):",A
Input "Imaginary part(b):",B
Gosub Close_window
'
Status%(19)=0  !Enable relevant options
Status%(20)=0
Status%(21)=0
'
Gosub Plot_set
Showm
Endif
Return
'
Procedure Picture_zoom
Local X0%,Y0%,X1%,Y1%,Mx%,My%,Newxmin,Newymin,Newxmax,Newymax
Gosub Bar_text("SELECT SECTION TO MAGNIFY")
Graphmode 3 !Set graphics mode to XOR so set not distorted
Showm
Repeat !
Until Mousek=1 And Mousey>10*Monitor%-2 !Wait for left button
'
X0%=Mousex
Y0%=Mousey
'
While Mousek=1
Mx%=Mousex
My%=Y0%+(Mx%-X0%)*0.59375
If Mx%>-1 And Mx%<320*Monitor% And My%>-1 And My%<200*Monitor% And
   Mx%<>X1% And Mx%>X0%
If Y1%>0
Box X0%,Y0%,X1%,Y1%
Endif
X1%=Mx%
Y1%=My%
Box X0%,Y0%,X1%,Y1%
Endif
Wend
Box X0%,Y0%,X1%,Y1%
'
Rem Calculate the variables for the new plot
If Mousek<>2 And Mousek<>3
Newxmin=Xmin+Increase*X0%
Newymin=Ymin+(200*Monitor%-Y1%-1)*Increase
```

Julia Sets

```
Newxmax=Xmin+Increase*X1%
Newymax=Ymin+(200*Monitor%-Y0%-1)*Increase
Max_iteration%=Int((Max_iteration%+((Xmax-Xmin)/(Newxmax-Newxmin)))
   /16)*16+17
Xmin=Newxmin
Ymin=Newymin
Xmax=Newxmax
Ymax=Newymax
'
Graphmode 0 !Put graphics mode back to normal
Gosub Plot_set
Endif
Return
'
Procedure Picture_parameters
Gosub Bar_text("PARAMETER DISPLAY")
Gosub Open_window("Current Parameters")
Print Using " Constant: ##.## +##.##i",A,B
Print Using "##.#### <x<##.####",Xmin,Xmax
Print Using "##.#### <y<##.####",Ymin,Ymax
Mi$=Str$(Max_iteration%)
Print "Max iteration:";
Form Input 4 As Mi$
Max_iteration%=Val(Mi$)
Gosub Close_window
Return
'
Procedure Colour  !This handles all three colour menu options
Local Choice%,Item%
Choice%=Menu(0)
For Item%=25 To 27
If Choice%=Item%
Status%(Item%)=1
Else
Status%(Item%)=0
Endif
Next Item%
Return
'
Procedure Dummy  !This should never be called
Local Button%
Gosub Bar_text("ERROR IN PROGRAM")
Alert 3,"Error!|The dummy procedure has|been called. Check the |On
   Menu(0) Gosub line.",1," Abort ",Button%
Edit
Return
'
Procedure Menu_status
Local Item%
If Monitor%=2
Status%(27)=2 !Disable colour option if necessary
Endif
```

```
For Item%=0 To 28
  Menu Item%,Status%(Item%)
Next Item%
Return
'
Procedure Bar_text(Message$)
  Print At(1,1);Spc(40*Monitor%);At((20*Monitor%-Len(Message$)/2),1);
     Message
Return
'
Procedure Plot_set
  Local P,Q,Temp_p,Button%,Iteration%
  Gosub Picture_parameters
  Gosub Bar_text("RIGHT HAND BUTTON HALTS PLOTTING")
  Status%(12)=0 !Remove tick from save option
  Hidem
  Increase=(Ymax-Ymin)/(190*Monitor%)
  For X=Xmin To Xmax Step Increase
    For Y=Ymin To Ymax Step Increase
      P=X
      Q=Y
      Iteration%=0
      Repeat
        Temp_p=P*P-Q*Q+A
        Q=2*P*Q+B
        P=Temp_p
        Inc Iteration%
      Until P*P+Q*Q>=4 Or Iteration%=Max_iteration%
      If Status%(25)=1 !Monochrome
        Color Iteration% Mod 2
      Endif
      If Status%(26)=1 !Set only
        Color Int(Iteration%/Max_iteration%)
      Endif
      If Status%(27)=1 !Full 16 colours
        Color Iteration% Mod 16
      Endif
      Plot Fn Round((X-Xmin)/Increase),(200*Monitor%-1)- Fn Round((Y-Ymin)/
         Increase)
      If Mousek=2
        Alert 3,"Are you sure you want|to stop drawing?",2, " Yes | No ",Button%
        If Button%=1
          Rem If the mouse key is pressed x and y are set their
          Rem maximum values, causing the FOR...NEXT loops to
          Rem think that they've finished
          X=Xmax
          Y=Ymax
        Endif
      Endif
    Next Y
  Next X
  Color 7 !Return drawing colour to normal
Return
'
```

Julia Sets

```
Procedure Check_res
Res%=Xbios(4)
If Res%=1 !Medium resolution not supported
  Alert 3,"This program only|works in high or|low resolution modes.",1,"
     Abort ",Dummy
  Edit !Quit back to editor (leave program if compiled)
Else
  Monitor%=Res%/2+1 !Note the use of integer monitor variable for speed
Endif
Ext$="PI"+Str$(Res%+1) !Set picture file extension
Return
'
Deffn Round(Float)=Int(Float)+Int(Frac(Float)*2)
'
Procedure Open_window(Title$)
Get 0,200*Monitor%-1,201,(155)*Monitor%,Store$ !Store backgrnd
Titlew 3,Title$ !Set title
Openw 3,200,(200-45)*Monitor% !Open it
Clearw 3 !Clear it
Return
'
Procedure Close_window
Closew 3 !Close window
Closew 0 !Return to full screen (0)
Put 0,(200-45)*Monitor%,Store$ !Replace window background
Return
'
Procedure Memory_initialise
Screen_info=Gemdos(72,L:34) !Reserve space for screen info blk
palette=Screen_info+2
Return
'
Procedure palette_store
Dpoke (Screen_info),Xbios(4) !Store screen resolution
'
For Register=0 To 15
  Dpoke palette+2*Register,Xbios(7,Register,-1) !Store palette
Next Register
Return
'
Procedure Memory_free
Status=Gemdos(73,L:Screen_info) !Free memory block
Return
```

Listing 5.2: The full Julia program

Enhancements to the Julia Program

Because there are so many similarities between the Julia and Mandelbrot sets, many of the ideas discussed at the end of Chapter 4 are also relevant here (3D landscape plotting and interior structure, for example). However, menu options for drawing Julia sets in these different ways would obviously be very useful. Alternatively, separate programs could be written to perform these tasks using information from files created using the original Julia program.

Altering the Program for Mandelbrot Exploration

The links to the Mandelbrot set also allow the program to be converted so that it draws the Mandelbrot set instead of, or as well as, Julia sets. This can very simply be achieved by altering the `plot_set` procedure because the size of the planes, and the method of colouring, is the same for both fractals.

A Resume Drawing Option

Despite the confirmatory alert box produced when a Julia plot is ended prematurely, it is feasible that plotting may be stopped accidentally. If this happens the Julia program is rather unforgiving – the only way to finish the plot is to draw it again completely from scratch. This problem could be overcome by storing the drawing position in a pair of global variables when the plot is interrupted and adding a *Resume drawing* menu option, which could be used to continue drawing from the stored position. If the interruption position was saved, as part of the Julia set files, it would also be possible to save an unfinished plot and complete it later – a particularly useful feature for performing long duration plots.

An Intelligent Search Path

A minor annoyance of the program is that when loading and saving Julia sets the search path given in the file selector is always `"A:\*.PI?"` even if, for example, the last file loaded was `"C:\CHAOS\PICTURES\JULIA.PI3"`. Most commercial programs would recognise that the user is storing all Julia sets in the PICTURES directory of the CHAOS directory on drive C and would use the search path `"C:\CHAOS\PICTURES\*.PI3"`. In the Julia program an *intelligent* search path of this sort could be maintained by taking the path part of the last filename selected and using this as the search path the next time the file selector command was used. Such an operation would be quite easy to perform using GFA BASIC, as it has powerful string handling capabilities. The basic process for determining the path from a filename is given below using the `"C:\CHAOS\PICTURES\JULIA.PI3"` example.

1. Extract from the filename all of the characters up to and including the last backslash (\), e.g. `"C:\CHAOS\PICTURES\"`

2. Add an asterisk character (*), e.g `"C:\CHAOS\PICTURES\*"`

3. Extract the last few characters from the filename, from the full stop character (.) onwards, and add this to the string to give the search path, e.g. `"C:\CHAOS\PICTURES\*.PI3"`.

Figure 5.9: Mathematical magnification of a section from Figure 5.7

Imitating Nature – Plants, Shrubs and Trees

Using chaos to draw abstract patterns such as the Mandelbrot set is certainly interesting, but many researchers are now turning their attention to the application of chaos techniques in mathematical generation of images depicting scenes from nature. It is relatively easy for an artist to copy such an image from the real world but generating a landscape, plant, animal or other natural object from scratch using nothing more than a set of rules and some mathematics is considerably more difficult. This challenging research not only produces pleasing pictures but also gives an insight into the processes which shape the world around us.

Ideally it should not be possible to tell the difference between a computer generated scene and a similar real life one. Due to the limited graphics facilities this ideal cannot be fully realised on the ST, but a useful insight into the relevant techniques and algorithms is given in this and the two following chapters concerned with imitating nature. This particular chapter deals with the creation of a variety of plants.

What is a Plant?

Although every plant species, and every individual within that species, is different it is easy to distinguish between plants and other objects just by looking at them. The way humans do this is to recognise a particular set of characteristics peculiar to plants. If the object we are trying to identify exhibits most of the characteristics from this set we can be fairly sure that it is a plant, even if we have never seen a plant of the same species before. To draw plants which are recognisable as such we need to know what these characteristics are, and how they can be re-created. This is not something specific to computer generated plants, artists also work in this way.

Imitating Nature - Plants, Shrubs and Trees 99

By briefly observing a variety of common plants a short list of features can be constructed which describes the silhouette of a typical plant. The list reads as follows:

❏ A single stem protrudes from the ground with many branches

❏ Most branches split into several smaller sub-branches until the smallest branches (e.g. twigs in trees) are reached

❏ Each plant has a large number of branch ends

❏ The nature of the branching is consistent throughout the plant.

Any biologist could easily discredit this definition, by quoting numerous examples which contradict it, but it serves as a good visual guide for the majority of plants. Note that flowers, fruit and leaves are not mentioned so as to keep the list valid for any time of the year and for as many plant types as possible.

Some of the characteristics typifying plants have been mentioned elsewhere in this book when describing fractals, the most obvious example being the branching of the Feigenbaum diagram. Because the branching is consistent throughout the length of the fractal the Feigenbaum diagram is said to be self-similar. Plants can also be shown to be self-similar using the same argument. A fern, for example, is composed of many miniature fern shapes, each closely resembling the plant of which it is part. Similarly the tiny veins on the underside of a leaf give a good approximation of the structure of the parent tree because, as stated above, the nature of the branching is the same throughout the tree.

Also like our computer generated fractals, the complex structure of a tree is created from a very simple process. Plants must have a large leaf-to-size ratio in order to trap as much light as possible for energy-gathering during photosynthesis. The best way for a plant to achieve this is to make every branch split into many sub-branches to create as many branch ends, hence as many leaves, as possible. This simple process is actually very successful in creating a very large surface area on which light can be trapped. To demonstrate the difference in surface area between a tree and a similar geometrical shape imagine painting a Christmas tree. You would need much more paint to cover its intricate structure of branches and needles than you would to cover a smooth faced cone of the same size.

In having such a large surface area trees are like the Mandelbrot and Julia sets, whose boundaries use up more space than expected for the number of dimensions in which they exist. By continuing this train of thought it is possible to give real plants, which exist in three dimensions, a fractal dimension of just over three. Because of this, and their self-similarity, plants can be classed as fractals, and by thinking of them in this context they can easily be drawn using similar methods to those employed in the fractal generators of earlier chapters. The exact set of rules, or algorithm, for drawing trees on a computer can be determined simply by observing the branching structure of plants in detail. Fractal plants, unlike previous fractals in this book, are created using a rule-based iterative process, rather than a mathematical one. This means that the process is easy to understand, but incorporating it into a program can be a challenge due to the reluctance of computers to deal in anything other than numbers.

Describing a Plant

The basic branching algorithm is relatively simple, but before considering this in depth we must devise a method of describing the structure of a plant which the ST can deal with. Not surprisingly GFA BASIC doesn't have built-in data structures for storing descriptions of trees! For demonstration purposes we can try to build up a description of the grass shown in Figure 6.1. This plant is referred to as a grass because it is simple in structure and only branches in one direction, making the principles easier to understand.

Figure 6.1: Example of a simple grass

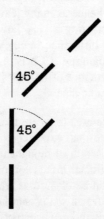

Figure 6.2: Segmented version of Figure 6.1 which the computer can deal with

Imitating Nature - Plants, Shrubs and Trees

The first thing that can be done to make the grass easier for the computer to handle is to split it up into straight segments of uniform length (see Figure 6.2). By doing this the branch lengths are said to be quantised, with the smallest possible branch being one segment long, and the length of every branch being a multiple of one segment length. Angles are also quantised for convenience, in steps of 45 degrees. Curved branches can be built up by putting several angled segments together. For more realistic, more detailed, plants shorter segments and smaller angles should be used, but because longer descriptions are needed to store such plants the values given above will be used in this initial discussion.

Simple notation can now be used to describe such plants using just three characters. The string shown below for the grass in Figure 6.2 will be used as the basis for the following discussion.

"1[1]1[11]"

This is a normal ASCII text string, not the most efficient method of storage as there are only three possible characters, but it is easy to deal with in GFA BASIC and makes future expansion simpler.

Each branch segment is represented in the string by the one character (1). The square brackets are collectively used to describe the plant's branches, where an open bracket ([) represents a 45 degree clockwise split from the current position (this usually represents the start of a branch) and a closed bracket (]) represents the end of a branch. Whole branches can easily be identified in such strings, as they are like miniature trees, with an equal number of open and closed brackets surrounding them. For example the [1] in the above description represents the first 45 degree branch.

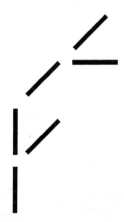

Figure 6.3: The grass with description "1[1]1[1[1]]"

Branches can also be given sub-branches. For example the longer branch ([11]) could be changed to [1[1]1], meaning that it had a single segment branching off at 45 degrees halfway along its length. The grass with this altered branch is shown in Figure 6.3.

Curved branches may also be described using this notation. The 180 degree curve in Figure 6.4 is described using the string:

"1[1[1[1[1]]]]"

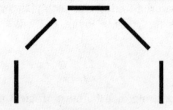

Figure 6.4: 180 degree curve in steps of 45 degrees

This is a particularly good example of how an open bracket does not necessarily represent the beginning of a new branch. The versatility of the notation means that it is possible to describe any plant which only branches in one direction, no matter how complex it is.

Turning Descriptions into Drawings

In order to write a BASIC program to turn such strings into pictures on the ST's screen we must define a method of interpreting each character into an action that the program can perform. To do this it is necessary to formulate a precise course of action for each of the three characters mentioned above. These actions are as follows:

1 Segment: Draw a uniform length line from the current position, at an angle equal to that held in the current angle count. The end of the new branch becomes the new graphics drawing position.

[45 degree clockwise split: Add 45 to current angle count.

] End of branch: Return to the beginning of the current branch (i.e. where the matching open bracket occurred) and decrease the angle count by 45 degrees.

The terms *angle count* and *drawing position* are best described using the turtle analogy on which the Logo programming language is based. Instead of the conventional approach of plotting points and lines at the graphics cursor, we imagine that a graphics *turtle* is being used. In this context a turtle is similar to a cursor in that

Imitating Nature - Plants, Shrubs and Trees

it can be placed at any pixel position on the screen, but in addition to a position the turtle also has an angular direction associated with it (the angle count). The direction is usually specified in degrees with zero being vertical and positive angles representing clockwise rotation from the vertical. Negative values are used to indicate anti-clockwise angles. Note that a line drawn from the current turtle position does not need to have its destination specified, as this is automatically calculated from the length of the line and the angle count. Figure 6.5 shows the path that a logo-type turtle would take when plotting the plant described by the example string introduced above.

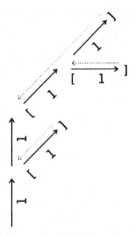

Figure 6.5: The path followed by a Logo turtle plotting the grass in Figure 6.3

The three actions have been incorporated into the draw_plant procedure in Listing 6.1, which is used to draw the grass from Figure 6.3. Note that GFA BASIC does not support turtle graphics so the program uses simple trigonometry to convert between the actions and the fdraw procedure. Note that draw_plant merely turns the string supplied in plant$ into a drawing, it doesn't actually generate plants.

The first section of the program initialises all global variables before calling the procedure to plot the plant. The variables are defined as follows:

plant$ Holds the string describing the grass, in plant generation programs later in this chapter the string will be created by the program.

unit_length Is the uniform segment length in low resolution pixels. This and unit_angle are variable so as to enable easy experimentation.

unit_angle Determines the angle at which branches split off. In this example it is 45 degrees, but as GFA BASIC only accepts radians it must be set to the equivalent radian value of PI/4 (see Appendix B for details of conversions).

angle_count Is used by the `draw_branch` procedure to store the current angle count, described above.

pointer Is used to store the current position in the string, i.e where to look for the next character to interpret. It is initially set to 0 because the first thing that the `draw_branch` procedure does is increment it (see below).

```
@Check_res
'
Plant$="1[1]1[1[1]1]" !Put sample string into plant
$Unit_length=10 !Set up constants for unit length...
Unit_angle=Pi/4 !...and unit angle
'
Gosub Draw_plant !Call procedure to draw the plant
@Waitmouse !Wait for a mouse key press
'
Procedure Draw_plant
Angle_count=0 !Initialise angle count
Pointer=0 !Set up initial string pointer position
Gosub Draw_branch(160,180)
Return
'
Procedure Draw_branch(X,Y)
Inc Pointer !Increment pointer position and...
Character$=Mid$(Plant$,Pointer,1) !...note character in that position
'
Repeat
If Character$="1" !If the character is a 1, draw a segment
@Fplot(X,Y)
X=X+Sin(Angle_count)*Unit_length
Y=Y-Cos(Angle_count)*Unit_length
@Fdraw(X,Y)
Endif
'
If Character$="[" !If it's a [ then branch off
Add Angle_count,Unit_angle
Inc Depth
Gosub Draw_branch(X,Y)
Endif
Inc Pointer !Note next character
Character$=Mid$(Plant$,Pointer,1)
Until Character$="]" Or Character$=""
'
Sub Angle_count,Unit_angle !If it's a ] (end of branch)
Return !decrease angle count and return
```

Listing 6.1: The draw_plant procedure with simple initialisation code

After initialisation the `draw_branch` procedure is called with the arguments 160 and 180. These are the *x* and *y* co-ordinates of the current drawing position that initially define the position of the bottom of the plant's stem.

The second section of the program is the actual definition of the `draw_branch` procedure. It is not immediately obvious how this operates, but all the techniques used are perfectly legal and can easily be understood if analysed logically. The first task of draw_branch is to increment the string pointer and copy the character at that new position in `plant$` into the `character$` variable. A REPEAT...UNTIL loop is then entered which checks for the open square bracket ([) and one (1) characters.

If the loop detects that the character is a 1 a segment of unit length is drawn at the current angle. This results in the end of the new line becoming the current drawing position. The next character is then extracted from the string, and if it is not a closed square bracket the loop then repeats.

If the character is not a 1, but an open square bracket, the `unit_angle` value is added to the current angle count and the `draw_branch` procedure is then called with the latest *x* and *y* arguments. If this happens the REPEAT...UNTIL loop does not continue until the `draw_branch` call returns.

If a closed bracket is taken from the string, or the end of the string is reached (`character$=""`), the REPEAT...UNTIL loop is terminated, `unit_angle` is subtracted from the angle count and the procedure returns to where it was called. In GFA BASIC the only time a procedure can return cleanly is at its end. This is obviously true for `draw_branch`. Usually the procedure will return to the open bracket clause of the IF...ENDIF construct. However, in the case where the end of the string has been reached the procedure will automatically return to where it was first called, at the head of the program, resulting in control being passed to the `waitmouse` procedure. This causes the program to pause until a mouse button is pressed, before terminating and returning control to the editing screen.

The method used here of calling a procedure from inside itself is known as *recursion* and, although a very powerful tool, it is sometimes hard to follow. Its use not only means that the same procedure can be used to draw all manner of branches, from twig to whole plant, but also means that we do not have to waste time and memory keeping track of where branching points occurred on the screen. Each time the `draw_branch` procedure invokes itself the previous group of arguments are stored by BASIC on a kind of internal stack. This means that as the procedure returns to where it was called the old values of *x* and *y* are taken off the stack and automatically become the current drawing position co-ordinates.

It is easy to experiment with different grasses by just altering the description string at the start of the program. Note that any valid string may be used but plants represented by particularly long strings may not fit on the screen. In this case you should reduce the `unit_length` and `unit_angle` constants as necessary. Some of the examples given later suggest suitable values for these constants which vary from the customary 10 pixel and 45 degree assignments. The most important thing to ensure, when trying to create a natural looking plant, is that each open bracket has a matching closed bracket.

String Generation

Interesting though the plant drawing program is, it will not actually generate plant descriptions. If you want a detailed piece of grass you have to type in a long string, which does not necessarily look very realistic. It is a fairly trivial matter to write a procedure to generate a plant's string description now that the necessary notational conventions have been established. Like other fractals, plants are generated using a simple, structure enriching, iterative process. The process used here is initially quite simple, but can be enhanced fairly easily. Basically every iteration sees each segment being replaced by a larger, more complex branch, thus making the plant larger while keeping the same level of relative complexity. This is achieved in the program by searching through the whole plant description (plant$), replacing all the 1 characters with a more complex user-defined string, set to determine the type of plant.

Listing 6.2 contains the replace_chars procedure to accomplish this task, along with the old plant drawing routine. In this example all 1s are replaced by the contents of one$, initially set to "11[1[1[1]]]", although any valid plant-type structure may be used. The plant description with which the program starts is the simplest possible structure, "1". It is important to note that there is no easy way to insert characters into a string, so what the program actually does is copy the contents of plant$ into the variable newplant$, character by character. However, if a one is found then the replacement string is copied across instead. Once the replacements are complete the contents of newplant$ are copied back to plant$ and newplant$ is cleared ready for the next iteration.

If you type in and run the program a tiny vertical sprig of grass will appear at the bottom of the screen. The mouse pointer will then appear, prompting you to press one of the mouse buttons. Upon pressing a button the first iteration of the string will be invoked and, after a short pause for calculation, the grass will begin to grow. The program then waits again for the mouse button before carrying out the next iteration. Figure 6.6, shows the appearance of the grass on successive iterations. Note that as the plant, and hence its string description, gets more complex the iterations take considerably longer to perform. If you have a compiler you may wish to use it here.

There are no limits imposed by the plant algorithms regarding the number of iterations performed but complexity is limited because GFA BASIC can only cope with strings of less than 32,768 characters. If the plant's description goes beyond this limit an error message will be displayed and the program will terminate. This is not as great a limitation as it seems because such complex plants could only be accommodated by the screen with a major loss of detail.

```
@Check_res
'
Plant$="1"
One$="11[1[1[1]]]"
Unit_length=2 !Set up unit length...
Unit_angle=Pi/16 !...and unit angle
'
Repeat
Cls
Gosub Draw_plant !Draw plant so far
Showm
@Waitmouse !Wait for mouse key
Hidem
Gosub Replace_chars !Make plant grow
Until Mousek=2
'
Procedure Draw_plant
Angle_count=0 !Initialise angle count
Pointer=0 !Set up initial string pointer position
Gosub Draw_branch(160,180)
Return
'
Procedure Draw_branch(X,Y)
Inc Pointer !Increment pointer position and...
Character$=Mid$(Plant$,Pointer,1) !...note character in that position
'
Repeat
If Character$="1" !If the character is a 1, draw a segment
@Fplot(X,Y)
X=X+Sin(Angle_count)*Unit_length
Y=Y-Cos(Angle_count)*Unit_length
@Fdraw(X,Y)
Endif
'
If Character$="[" !If it's a [ then branch off
Add Angle_count,Unit_angle
Gosub Draw_branch(X,Y)
Endif
Inc Pointer !Note next character
Character$=Mid$(Plant$,Pointer,1)
Until Character$="]" Or Character$=""
'
Sub Angle_count,Unit_angle !If it's a ] (end of branch) then
Return !decrease angle count and return
'
Procedure Replace_chars
Pointer=0 !Set initial pointer position
Repeat
Inc Pointer !Increment pointer position
Character$=Mid$(Plant$,Pointer,1)!Extract character
If Character$="1" !If it's a 1 then replace it with one$
Newplant$=Newplant$+One$
Else
Newplant$=Newplant$+Character$ !Otherwise leave it unchanged
Endif
Until Pointer=Len(Plant$)
Plant$=Newplant$ !Replace old plant$ with new one
Newplant$="" !Clear newplant$ just in case
Return
```

Listing 6.2: The replace_chars and draw_chars procedure being used to generate, and then draw a plant

Figure 6.6 (a): The appearance of the grass with replacement string "11[1[1[1]]]" after four iterations...

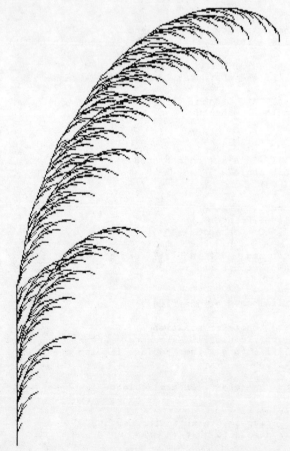

Figure 6.6 (b): ...and after six iterations

The program produces some quite interesting, sometimes unexpected, results making experimentation very worthwhile. Table 6.1 shows some notable values for unit_angle, unit_length, plant$ and one$. When using your own values be sure to have matching brackets, and also try to plan ahead so that the tree gets larger as well as getting more complex. For example a replacement for one of "11[1]" will cause the replaced branch to get longer whereas "1[1]" will just change it into a two branches, each similar in length to the original.

Table 6.1: Variable assignments for a variety of grasses

plant$	one$	unit_length	unit_angle
1	11[11[[1]]]	3	PI/12
1	1[1]1[[1[[1]]1[[1]]]]	4	PI/16
1	11[1[[1[1[[1]]]]]]	4	PI/16

Trees

Only grasses have been drawn thus far because their branches split off in a single direction. Now that the basic algorithm has been established it is fairly easy to alter it for the bi-directional type of branching found in trees and shrubs. All that needs to be done is to add two new characters, the curly brackets ({ and }), to the set recognised by the draw_branch plant visualisation procedure. The curly brackets are used to enclose descriptions of branches which split off anti-clockwise. The relevant actions for each curly bracket, shown below, are very similar to those associated with the equivalent square brackets.

{ **Anti-clockwise split:** Subtract unit_angle from current angle count.

} **End of branch:** Return the graphics cursor to the beginning of the current branch (i.e. where the matching open bracket occurred) and increase the angle count by unit_angle.

Listing 6.3 shows the adapted definition of the draw_branch procedure required for bi-directional branching. The program code used to check for and act on curly brackets is very similar to that for square brackets. There is no need to edit the definition of replace_chars because brackets are not replaced under any circumstances.

If you have changed the draw-branch procedure, but haven't altered the definition of plant$ and one$ at the start of the program you will see exactly the same output as before. However, the whole program can now cope with curly brackets and if the variables are set to some of the values from Table 6.2 trees with bi-directional branching will be generated. Note that it is perfectly natural for anti-clockwise branches to have clockwise sub-branches and vice-versa. Some example trees are shown in Figures 6.7 and 6.8.

```
Procedure Draw_branch(X,Y)
Inc Pointer !Increment pointer position and...
Character$=Mid$(Plant$,Pointer,1) !...note character in that position
'
Repeat
If Character$="1" !If the character is a 1, draw a segment
@Fplot(X,Y)
X=X+Sin(Angle_count)*Unit_length
Y=Y-Cos(Angle_count)*Unit_length
@Fdraw(X,Y)
Endif
'
If Character$="[" !If it's a [ then branch off clockwise
Add Angle_count,Unit_angle
Gosub Draw_branch(X,Y)
Endif
If Character$="{" !If it's a { then branch off anti-clockwise
Sub Angle_count,Unit_angle
Gosub Draw_branch(X,Y)
Endif
Inc Pointer !Note next character
Character$=Mid$(Plant$,Pointer,1)
Until Character$="]" Or Character$="}" Or Character$=""
'
If Character$="]" !If it's a ] (end of branch) then
Sub Angle_count,Unit_angle !decrease the angle count..
Endif
If Character$="}" !If it's a } then increase angle...
Add Angle_count,Unit_angle
Endif
Return !...and return from the procedure
```

Listing 6.3: The adapted definition of draw_branch

Table 6.2: Variable assignments for a variety of trees

plant$	one$	unit_length	unit_angle
1	11{1{1}[1]}[1{1{1}}]]	3	PI/8
1	11[1[[1[1[[1]]]]]]{1{{1{1{{1}}}}}}	4	PI/16
1	11{1{1}1}{11[1]1}	3	PI/8
1	1{{11}}[[11]]1[{{11}}[[11]]1{{{1}}[[1]]1]]	8	PI/10

Imitating Nature - Plants, Shrubs and Trees 111

When is a Plant Not a Plant?

The plant generation program developed in this chapter can also be used to create a wide variety of non-biological structures. For example, by replacing the variable assignment lines with those shown below it is possible to produce a very complex snowflake structure (see Figure 6.9).

Figure 6.7: Tree produced using the fourth set of values from Table 6.2

112 Computers and Chaos: Atari ST

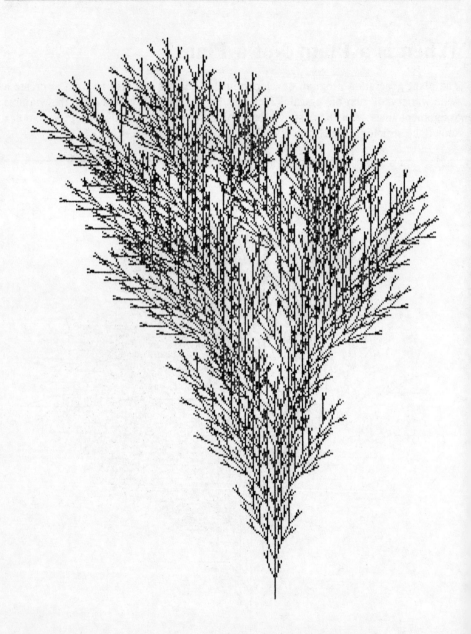

Figure 6.8: Tree produced using the third set of values from Table 6.2

Naturally the character replacement technique is exactly the same as that used for the plants above, and even the replacement string looks familiar, but in this case the plant initially has six branches, each one at 60 degrees to its neighbour. In this example one$ specifies that each segment should be replaced by a slightly longer segment

Imitating Nature - Plants, Shrubs and Trees

with a bifurcation at the end. Note that because the strings include curly brackets they can only be implemented in the bi-directional plant drawing program.

```
plant$="{{{11}}}{{11}}{11}[11][[11]]11"
one$="11[1]{1}"
unit_length=1
unit_angle=PI/3
```

Figure 6.9: Snowflake generated using the plant program

In addition the origin of the plant must be moved from the bottom of the screen to the middle, because the flake grows in all directions. The plant's origin is changed by altering the line which calls the `draw_branch` procedure so that it reads:

```
Gosub Draw_branch(160,100)
```

which causes a different initial drawing position to be sent to the procedure.

It is accepted that every snowflake should be unique in structure. This is hard to achieve on a computer but the structure of the one in the plant program can be drastically altered by simply changing the contents of one$. As an example, "11[1]{1}1" would give trifurcations at the end of each branch.

The C-Curve

If you are familiar with chaos theory you may be able to identify other, non-branching applications of the character replacement routine used in the plant generation program. Some of the earliest fractals were based on the replacement of straight line sections with more complex line structures, something which the plant program excels at. These were discovered first because good approximations of them can easily be drawn by hand, without the need for excessive number crunching.

One such fractal is the C-curve, shown in Figure 6.10. The basic algorithm for drawing it is to start with a single line and then repeatedly replace every straight line with two sides of a right angled triangle. The effect of the first few iterations is shown in Figure 6.11.

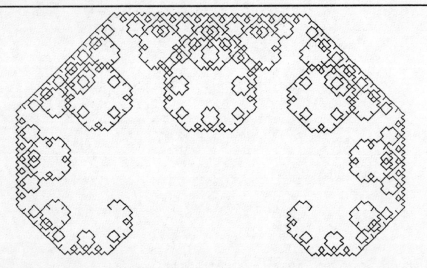

Figure 6.10: The C-Curve after 11 iterations

After several iterations the curve becomes very complex in structure, resembling an elaborate letter C, hence the name. The relevant modifications to the program are:

```
plant$="[[1]]"
one$="{1[[1{"
unit_length=3
unit_angle=PI/4
Gosub Draw_branch(100,160)
```

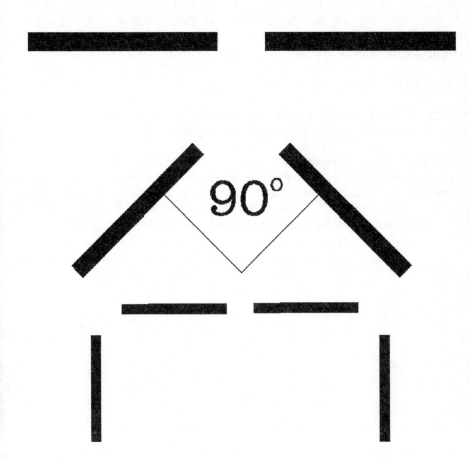

Figure 6.11 (a) – (d): The replacement process used to create the C-curve

These values have been specially formulated to give the most detailed and largest C-curve possible. In this case the complexity of the fractal is limited by the number of recursive procedure calls that GFA BASIC can cope with, rather than the length of the description string. The recursion problem becomes particularly acute in the C-curve because the replacement string (one$) contains unmatched open brackets. This means that the draw_branch procedure has to call itself many more times than usual (once for every bracket), so memory is rapidly used up to store the ever-growing stack of arguments from previous calls.

Because the problem originates from excessive recursion the *Memory full* error box will appear while drawing is taking place, rather than during string processing. This is how this type of error can be distinguished from one caused by plant$ becoming too long. Note that for this particular fractal (and the Koch curve below) a non-recursive drawing procedure would be preferable as it would facilitate the creation of more detailed images.

The Koch Curve

The Koch curve is a kind of aesthetically pleasing version of the C-curve, which originates from an equilateral triangle (see Figure 6.12).

The middle third of each side of the triangle is replaced by two sides of a smaller equilateral triangle, leaving a 12-sided shape. The same replacement process is then repeated on every side of the new shape, and on every side of the subsequent shape. After an infinite number of iterations the shape becomes the Koch curve (Figure 6.13), often referred to as the Koch flake due to its snowflake-like structure.

Figure 6.12 (a) - (c): The replacement process used to create the Koch curve

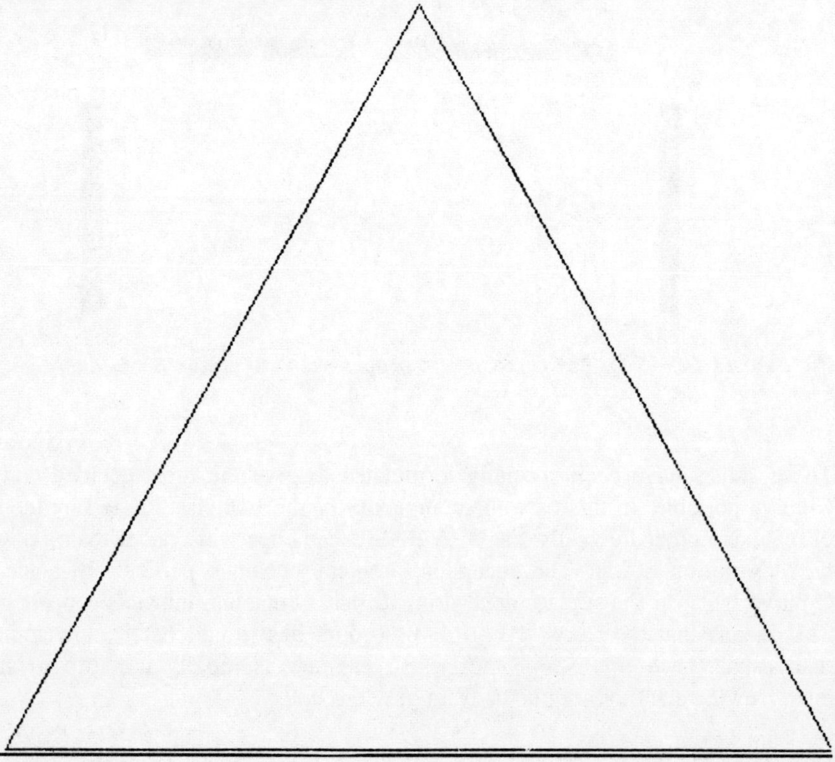

(a).

Imitating Nature - Plants, Shrubs and Trees   117

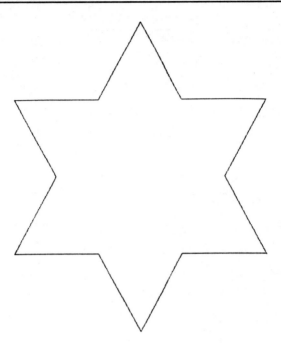

Figure 6.12 (b)

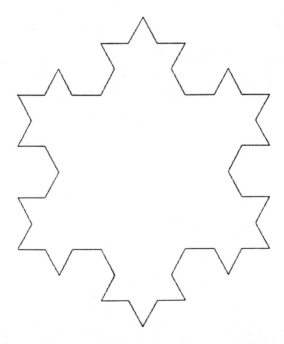

Figure 6.12 (c)

Figure 6.13: Approximation of the Koch curve

The replacement lines needed to produce the curve using the plant program are:

```
plant$="1[[1[[1]]]]"
one$="1{1[[1{1"
unit_length=200
unit_angle=PI/3
Gosub Draw_branch(120,200)
```

To make the Koch curve appear to change, rather than grow, the unit_length variable must initially be large and then be reduced to a third of its value after each iteration. This is done by altering the first REPEAT...UNTIL loop in the plant generation program so that it reads as follows:

```
Repeat
Cls
Gosub Draw_plant !Draw plant so far
Showm
@Waitmouse !Wait for mouse key
Hidem
Gosub Replace_chars !Make plant grow
Unit_length=Unit_length/3 !Reduce size by a third
Until Mousek=2
```

Of course, the curve created here is only an approximation of the real Koch curve due to the limitations of the computer, which make it impossible to perform an infinite number of iterations.

Like many other fractals in this book the Koch flake has an infinite length. Although this is impossible to prove practically it can be done using elementary maths. Assuming that the length of each side of the initial triangle is one metre, its total perimeter is three metres, as shown in Figure 6.14. After one iteration the length of each side becomes $1^1/_3$ (4/3) metres long due to the introduction of the triangular deviation, giving a perimeter of 3*(4/3) = 4 metres (an increase of one third).

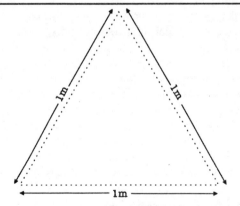

Figure 6.14: The size of the equilateral triangle from which the Koch curve is produced

On the next iteration each side is only a third of a metre long, and is only extended by a ninth of a metre to 4/9 metre. But because there are 12 sides the overall length increases by a third to 12*(4/9) = $5^1/_3$ metres. On each iteration the length of the perimeter increases by a third (represented mathematically as multiplying by 4/3), so on successive iterations the perimeter gets larger at an increasing rate. For example, after two iterations the perimeter would be:

3 * 4/3 * 4/3 = 5.33m

After four iterations the perimeter would be:

3 * 4/3 * 4/3 * 4/3 * 4/3 = 9.48m

As a general rule the perimeter length can be expressed as:

length = $3 * (4/3)^n$

where n is the number of iterations. Using this expression it is possible to see that the length quickly becomes very large. After only 45 iterations ($n=45$) it is already over one million metres. Substituting infinity for the iteration number shows that at this stage the perimeter is itself infinite in length. However, the area of the shape stays small implying that it is possible to store an infinitely long line in a finite space (less than one square metre in this example). To reflect the way that the Koch curve is more like a two dimensional shape than a one dimensional line (in its ability to fill space) it is given a fractal dimension of 1.2818.

The infinite complexity of the Koch curve is very difficult to comprehend. Magnifying one section reveals that it is made up of similar, equally detailed sections, and further magnification reveals even smaller and more detailed sections. Unlike normal geometrical shapes it would be impossible to magnify the curve so as to expose the straight lines from which it is totally composed, a concept familiar from earlier chapters.

In classical mathematics the slope (or gradient) of a curve at any point can normally be found by mathematically 'magnifying' it until it approximates to a straight line, for which the gradient can easily be determined. However, because the Koch curve changes direction at every point and never approximates to a straight line it is impossible to find a single value for the gradient at any given point. This is just one of many reasons why classical mathematicians have always viewed the Koch curve (and similar fractals) with a noticeable amount of disdain.

Further Experimentation

Growing artificial plants is a very interesting pastime, so I have attempted to write the accompanying program in such a way as to make experimentation in this area as easy as possible. The way the fractal grows can be altered simply by changing the few variables initialised at the start of the program. Many other natural fractal patterns such as lightning and frost on a window can be created using the methods and programs discussed in this chapter. A variety of more specific enhancements are outlined below.

Branches with Variable Thickness

One of the most noticeable simplifications of the current model is that the thickness of each branch is in no way related to the load it has to bear. Although grasses have an acceptable appearance, trees look as though their trunks are as thin as twigs. The GFA BASIC DEFLINE command allows us to set the line width, which is then used in all future calls to DRAW, CIRCLE and the other line drawing commands. By reducing the line width in this way whenever a new branch begins (whenever an open bracket is reached) the branches realistically appear to get thinner towards their ends.

More Complex Complexity

The rule set used throughout this chapter only allows for one type of segment character, "1". The addition of a second type of segment, which has a different replacement string, means that still more complex and diverse branching effects can be achieved using the existing algorithm. The character we shall use for this other type of segment is the zero (0). The changes required to implement the new segment type are fairly minor. The first is near the beginning of the draw_branch procedure. This line:

```
If Character$="1"
```

should be changed to:

```
If Character$="1" Or Character$="0"
```

so that the program draws a segment if a 1, or a zero, is found. The replace_chars procedure should then be edited to read:

```
Procedure Replace_chars
Pointer=0 !Set initial pointer position
Repeat
Inc Pointer !Increment pointer position
Character$=Mid$(Plant$,Pointer,1) !Extract character
If Character$="1" !If it's a 1 then replace it with one$
Newplant$=Newplant$+One$
Endif
If Character$="0" !If it's a 1 then replace it with one$
Newplant$=Newplant$+One$
Endif
If Character$="{" Or Character$="}" Or Character$="[" Or Character$="]"
Newplant$=Newplant$+Character$ !Otherwise leave it unchanged
Endif
Until Pointer=Len(Plant$)
Plant$=Newplant$ !Replace old plant$ with new one
Newplant$="" !Clear newplant$ just in case
Return
```

Changes are needed here because it can no longer be assumed that if the character taken from the string is not a 1 it will not need replacing – each character must be explicitly checked for. If you are attempting to implement this program in C it would be more efficient to use SWITCH and CASE here.

This modified program will still draw all the fractals discussed earlier in the chapter, but just as all occurrences of "1" are replaced by the contents of one$, all zeros are replaced by the contents of zero$. The example variable assignments below will allow you to pilot the new program.

```
plant$="10"
one$="10[01]{10}"
zero$="1[1{01}0]{11}"
unit_length=5
unit_angle=PI/6
```

Note that any number of different segment types may be added to the program in the same way. If there are more than two possible types of segment it is advisable to use an array to store the relevant replacement strings, failure to do this will make the replace_chars procedure very much longer and will result in slower execution.

Using C

The tree drawing routine is a very good example of a program that should have been written in C. The messy global variables and constants would not be a problem in C as values can easily be returned at any point in a routine and true constants are well supported. Although GFA BASIC allows functions to return values, its functions are limited to one line, which is too short in this case. C also makes available more efficient, user-definable data structures in which plants could be stored.

7

Imitating Nature – Fractal Landscapes

Probably the slowest natural process discussed in this book is landscape dynamics. Although normally appearing stationary for many thousands of years, the Earth's landscape is actually undergoing constant change due to erosion, volcanic activity and more gradual land movements. The lack of speed makes the process slightly easier to predict than something like the weather, but no easier to simulate because it is still inherently chaotic. This chapter describes some of the simplest methods for creating natural looking landscapes and proceeds to discuss the application of landscape drawing techniques to fractals such as the Feigenbaum diagram and the Mandelbrot set. Landscape generation programs produce aesthetically pleasing results and provide an interesting introduction to the processes which create terrestrial landscapes, making experimentation very worthwhile.

Something common to all landscapes is that they are three dimensional, they stretch out horizontally in two directions and also vary in height. To draw such objects it is obviously necessary to find a technique which will allow plotting in three dimensions. The most popular way of doing this is called isometric drawing.

Isometric Drawing

On a two dimensional plane, such as that used to display most graphs, there are two directions of movement, represented by two axes at right angles to one another, as shown in Figure 7.1 (overleaf). The two directions of movement are vertical (represented by the y axis) and horizontal (represented by the x axis).

In three dimensional space, however, there is an additional direction of movement, and hence three axes (x, y and z). As the screen of the ST is only two dimensional it is of course impossible to display three dimensional objects, but take a look at Figure 7.2 (overleaf).

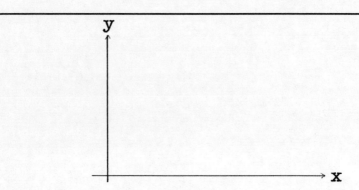

Figure 7.1: Typical two dimensional plane

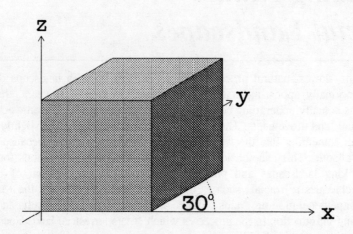

Figure 7.2: Isometrically drawn three dimensional space and cube

This shows the three axes used to represent three dimensional space, with a cube placed where they meet. Obviously this diagram is not really three dimensional, but the illusion of three dimensions is produced sufficiently to convey the fact that the object shown is a 3D cube. This effect is achieved by plotting the x and z axis as normal but drawing the y axis at 30 degrees to the x axis, and treating all corresponding movements in the y direction in the same way. The technique is called isometric drawing and is frequently used by draughtsmen when producing illustrations of their designs, and can also be found in some computer games and demos. Drawing objects such as spheres and cubes isometrically can be a fairly complicated process as a method must be devised for storing the structure of objects, but plotting landscapes is somewhat easier due to landscapes simply being uneven planes.

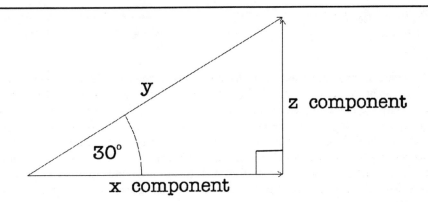

Figure 7.3: Mathematical splitting of y into horizontal and vertical components

To perform isometric drawing in GFA BASIC an algorithm is required to convert a three dimensional (x,y,z) co-ordinate to the two dimensional form accepted by the BASIC PLOT and DRAW functions. This algorithm involves splitting the isometrically drawn y position into its constituent horizontal and vertical components, as shown in Figure 7.3, and adding these components to the x and z positions for the point. By merging y with x and z we have effectively eliminated y and are now left with a two dimensional co-ordinate of the form (x,z) which can be used directly with PLOT, POLYLINE or any other graphics function available from BASIC.

If the components of y are to be added to x and z to produce a composite co-ordinate it is necessary to determine the size of the components. This is done by calculating the ratio between them using elementary trigonometry, and since it is a ratio, rather than an absolute value, it is the same for any value of y. The mathematical tangent function can be used to find this universal ratio because the tangent of an angle is defined as the ratio of the length of the angle's opposite and adjacent sides in a right angled triangle. From Figure 7.3 it is clear that the ratio of the z component to the x component is the tangent of 30 degrees. The value of this tangent can be found using the tan button of a pocket calculator, which should return a result of approximately 0.6. The mathematical interpretation of this process is shown below (angles, right angled triangles and the tangent function are discussed in detail in Appendix B).

$$\tan 30 = \frac{z \text{ component}}{x \text{ component}} = \frac{6}{10} \quad (\text{or } 0.6 \text{ or } 6{:}10)$$

Now that the ratio of the z component to the x component is known it is possible to plot points on the screen isometrically. The method for doing this is best conveyed by the example plane drawing program in Listing 7.1. Here a pair of FOR...NEXT loops are used to completely traverse a 48x48 point plane, on which each point is plotted isometrically. The (x,y) position of the point passed to the fplot procedure is calculated as follows:

Position passed to fplot	Offset	x and z parts	y components
horizontal	=	$x*5 +$	$1.7*y$
vertical	=	$148 -$	y

There are several important points to note in these equations. Firstly the z position of every point is zero, so this is not included in the calculation of the y position. Secondly the horizontal component of y is equal to $1.7*y$ and the vertical component is equal to y (or $1*y$), so the ratio is correct for 30 degrees ($1/1.7 = 0.6$). Also note that the vertical position passed to fplot is inverted by subtracting it from 148 – necessary because of the ST's upside-down vertical axis. The x position is multiplied by 5 to ensure that the maximum screen area is utilised, although in this example the fact that z is equal to zero means that there are large spaces at the top and bottom of the screen.

```
@Check_res
'
For Y=47 To 0 Step -1
For X=0 To 47
@Fplot(X*5+Y*1.7,148-Y)
Next X
Next Y
@Waitmouse
```

Listing 7.1: The plane drawing program

The plane produced by Listing 7.1 (in high resolution mode) is shown in Figure 7.4(a). A landscape could easily be represented here by giving each point a height determined by some particular method, but since each point is plotted as a single pixel, with no connections to adjacent pixels the points would quickly become confused making the image meaningless.

Figure 7.4(b) shows the single pixel plotting method being used to draw a landscape 'testcard'. A more suitable plotting method involves joining together all points that have the same y position in order to create a series of 48 lines depicting the plane.

As Figure 7.4(c) shows this is a great improvement, but the plane appears transparent in places because lines which are meant to be behind or below other parts of the plane can still be seen. The process of removing unwanted lines to make an object appear solid is called 'hidden line removal', the result of which is shown in Figure 7.4(d).

The best way to remove hidden lines in our case is by using the GFA BASIC POLYFILL command to draw each line and fill the gap between it and the last line drawn, in the background colour. Figure 7.5 shows the way in which a polygon is constructed from the two most recent lines so that the gap between them is totally enclosed and can be automatically filled in.

Imitating Nature - Fractal Landscapes

Figure 7.4(a): Isometric 3D plane produced by Listing 7.1

Figure 7.4(b): Testcard landscape drawn using single pixels

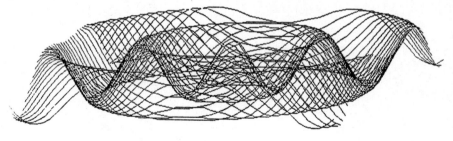

Figure 7.4(c): Testcard consisting of lines

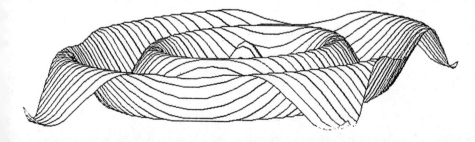

Figure 7.4(d): Testcard drawn using lines and hidden line removal

The programming needed to implement this hidden line removal system is beyond the scope of this book, and unrelated to chaos theory, so an 'off-the-shelf' procedure has been included in Listing 7.2 which draws a plane with hidden lines removed, from the data describing the height of each point stored in the 48x48 element z% array. The size and position of this plane, and the relationship between z% co-ordinates and their screen positions are shown in Figures 7.6 and 7.7. This procedure definition will need to be included in all programs given later in this chapter which contain calls to landscape.

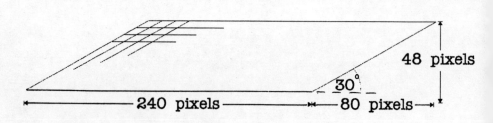

Figure 7.5 Application of filled polygons to remove hidden lines

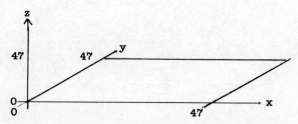

Figure 7.6: Size of the plane produced by the landscape procedure (in low-resolution pixel units)

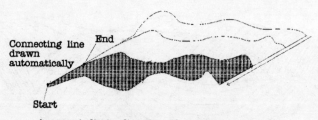

Figure 7.7: Relationship between elements of the z% array and the plane drawn by landscape

Imitating Nature - Fractal Landscapes

```
Procedure Landscape
Cls
Deffill 1,0
Hidem
'
For Y=47 To 0 Step -1
For X=0 To 47
Poly_x%(X)=X*5*Monitor
Poly_y%(X)=-Z%(X,Y)*Monitor
Next X
If Y<47
For X=0 To 47
Poly_x%(95-X)=X*5*Monitor+1.7*Monitor
Poly_y%(95-X)=-Z%(X,Y+1)*Monitor-Monitor
Next X
Polyfill 96,Poly_x%(),Poly_y%() Offset Y*1.7*Monitor,(148-Y)*Monitor
Rem Enhancements may be added here (see text)
Else
Polyline 48,Poly_x%(),Poly_y%() Offset Y*1.7*Monitor, (148-Y)*Monitor
Endif
Next Y
Return
```

Listing 7.2: The landscape procedure

Using the Landscape Procedure

To use this procedure it is necessary to dimension two 96 element temporary storage arrays – `poly_x%` and `poly_y%` and also the 48 by 48 element $z\%$ array. The height of each of the 2304 (48*48) points on the plane should then be placed in the correct elements of the $z\%$ array before `landscape` is called to plot the plane. As an example the program shown in Listing 7.3 would be used to set the height of the point in the middle of the plane to be 20 (low resolution) pixels high and draw the resulting landscape. Running this short program will determine whether the landscape procedure has been entered correctly, and also demonstrates the minimum program construction needed to use `landscape`.

```
@Check_res
'
Rem Set up arrays
Dim Poly_x%(95)
Dim Poly_y%(95)
Dim Z%(47,47)
'
Rem Set height of point
z%(23,23)=20
'
Rem Draw landscape
Gosub Landscape
@waitmouse
```

Listing 7.3: Setting a single height point with the aid of landscape

Naturally it would be impractical to set the height of all 2304 points in this way, so a nested pair of FOR...NEXT loops are generally used to fill the array instead. An example of a program using such a method, to draw the testcard, is shown in Listing 7.4. Note that the testcard is actually a three dimensional cosine curve on which the height of each point is determined according to its distance from the centre of the plane, and is in no way related to fractal techniques. Fractal landscapes are discussed in the next section.

```
@Check_res
'
Rem Set up arrays
Dim Z%(47,47)
Dim Poly_x%(95)
Dim Poly_y%(95)
'
Rem Set point heights in z% array
Print "Calculating..."
'
For X=0 To 47
For Y=0 To 47
Z%(X,Y)=Cos(Sqr(((24-X)^2)+((24-Y)^2))/2)*20
Next Y
Next X
'
Rem Draw landscape
Gosub Landscape
@Waitmouse
```

Listing 7.4: Using a FOR...NEXT loop to set the height of all the points

Improving the Landscape Procedure

The planes produced by landscape can be made to look more solid by joining points along the direction of the y axis to make the plane look like it is composed of tiny quadrilaterals, as shown in Figure 7.8 (overleaf).

The landscape procedure can be adapted for this purpose by adding the following lines to the procedure definition given in Listing 7.2, at the point marked for enhancements.

```
For X=0 To 47
Draw Poly_x%(X)+Y*1.7*Monitor,Poly_y%(X)+(148-Y)*Monitor To
         Poly_x%(95-X)+Y*1.7*Monitor,Poly_y%(95-X)+(148-Y)*Monitor
Next X
```

Other improvements which could be included in this section of the routine, such as the addition of shading, are discussed at the end of this chapter.

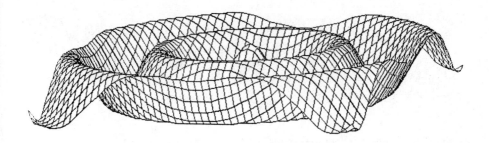

Figure 7.8: Testcard landscape composed of quadrilateral segments

Generating Pseudo-Natural Landscapes

Now that we have a method of plotting a landscape we can begin to use chaos techniques to determine the height of each point on the plane, in order to give the landscape a more natural appearance.

There are numerous methods available for producing landscapes with features similar to those found in the real world, and more realistic ones are being produced all the time. Upon looking at a natural landscape it is evident that the process used to set the height of each point on the plane would have to be random, and a first attempt at a landscape program may look like Listing 7.5.

```
@Check_res
'
Dim Z%(47,47)
Dim Poly_x%(95)
Dim Poly_y%(95)
'
Print "Calculating..."
'
For X=0 To 47
For Y=0 To 47
Z%(X,Y)=Random(21)-10 !Set height of point
Next Y
Next X
'
Gosub Landscape
@Waitmouse
```

Listing 7.5: A totally random landscape program

This sets the height of each point randomly within the range of −10 to 10, and certainly does not produce a natural landscape, as Figure 7.9 demonstrates.

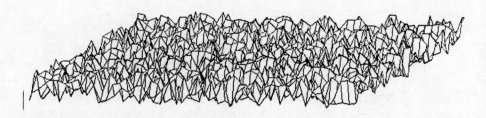

Figure 7.9: Totally random landscape

The problem is that the degree of randomness is too great, causing incredibly steep slopes and sudden points of height to be produced. On a real landscape such steep slopes would be made gentler by a gradual process of weathering. It is true that most landscape generators rely on some kind of random process, but to generate a realistic landscape it is necessary to control the randomness in some way so that it does not get too out of hand, as in the Sierpiński triangle. Two generators based on controlled randomness are presented below.

Inheritance

The main problem with Listing 7.5 is that there is no continuity between adjacent points, meaning that huge unnatural pillars of land can stand out of the ground. Using a process of inheritance the height of each point is always related in some way to the height of at least one neighbouring point. A simple example of this can be demonstrated in two dimensions using the program in Listing 7.6.

```
@Check_res
'
Y=0 !Set initial y position
@Fplot(10,100-Y) !Plot the position
For X=11 To 310
Y=Y+Random(5)-2 !Calculate new y value
@Fdraw(X,100-Y) !Draw line to new point
Next X
@Waitmouse
```

Listing 7.6: Relating the height of a point to its neighbour using random inheritance

In this example the first point (at the left of the screen) is set to be at a height of zero pixels. A FOR...NEXT loop then begins to draw the rest of the points, whose heights are calculated by adding a random number, between –2 and 2, to the height of the previous point. The result, shown in Figure 7.10, is reminiscent of the cross section of a rather mountainous landscape.

Imitating Nature - Fractal Landscapes

Figure 7.10: Random inheritance in two dimensions

The roughness of the terrain can easily be altered by changing the range of the random number added to the previous point, for instance the following line will give gentler slopes:

```
Y=Y+Random(3)-1
```

By flattening points below a certain height, referred to as the sea level, it is possible to create the effect of lakes, rivers and seas between pieces of land, as shown in Figure 7.11.

Figure 7.11: Cross-section of a landscape with seas and lakes

Replacing the line which draws between points with the IF...ENDIF construct below will draw all points with y positions below zero as if their y positions were zero. In this example the sea level can be said to be zero. Note that the value of *y* is not actually altered, as this would upset the inheritance process.

```
If Y>0 !If y is above sea level
@Fdraw(X,100-Y) !Draw line to new point
Else
@Fdraw(X,100) !Draw line at sea level
Endif
```

This method of flattening sub-zero sections of the plane to create areas of water can be used in all of the natural landscape generators detailed in this chapter.

Enhancing the inheritance process for three dimensions is relatively easy, since the plane plotting program has already been provided. However, in three dimensions, the inheritance pattern is slightly different because each point has up to two adjacent points from which to inherit values, as shown in Figure 7.12.

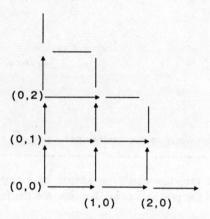

Figure 7.12: Inheritance on a plane in three dimensions

It can be seen from this diagram that the first point calculated, at (0,0), is completely random because there are no adjacent points from which heights can be inherited. Other points along the left and lower sides of the plane have only one adjacent point whose height has been calculated, so they inherit their height from a single point only, as in the two dimensional example in Listing 7.6. All other points on the plane have two processed neighbours, so the heights of the two adjacent points are combined, by adding them together and dividing the result by 2. The resulting value is used to determine the height of the new point. Listing 7.7 uses this method to fill the $z\%$ array with values and will plot the result.

As the program shows, the very artificial inheritance method of determining the height of the points produces surprisingly realistic landscapes, an example of which is shown in Figure 7.13.

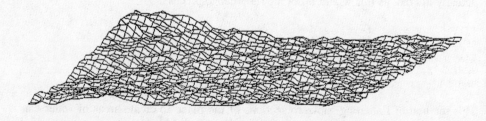

Figure 7.13: Pseudo-natural landscape created by Listing 7.7

A more refined type of inheritance can be created by basing the change in height between one point and its neighbour on the height difference between the neighbour and its other neighbour. This creates slightly more natural landscapes with gentler slopes.

```
@Check_res
'
Dim Z%(47,47)
Dim Poly_x%(95)
Dim Poly_y%(95)
'
Print "Calculating..."
'
For X=0 To 47
For Y=0 To 47
If X=0 And Y=0 !(0,0) - no inheritance
Z%(X,Y)=random(5)
Endif
If X=0 And Y>0 !Left edge of plane - one neigbour
Z%(X,Y)=Z%(X,Y-1)+Random(5)-2
Endif
If X>0 And Y=0 !Nearside edge - one neighbour
Z%(X,Y)=Z%(X-1,Y)+Random(5)-2
Endif
If X>0 And Y>0 !Rest of plane - two neighbours
Z%(X,Y)=(Z%(X,Y-1)+Z%(X-1,Y))/2+Random(5)-2
Endif
Next Y
Next X
'
Gosub Landscape
@Waitmouse
```

Listing 7.7: *Using neighbours to determine the height of the new point*

Faulting

In the real world landscapes are formed as the result of a variety of processes, including large movements of land, one type of which forms the basis of the faulting method of landscape generation. Geologists have found that the Earth's crust can be thought of as a relatively thin layer of solid land floating on a sea of molten rock, just like ice floating on water. Because of pressure points and other phenomena occurring inside the Earth it is never possible for the crust to be in one piece. Instead it is made up of enormous plates of land, which overlap and slide underneath each other where they meet. When the edges of two such plates push together, but do not slide over each other, a large force is exerted between them.

Initially this force goes unnoticed, but it eventually builds up to such an extent that one of the plates must suddenly rise or fall in order to relieve the pressure. This sudden movement of land is called a fault, shown pictorially in Figure 7.14. After hundreds of faults occurring over millions of years spectacular mountainous landscapes can be created.

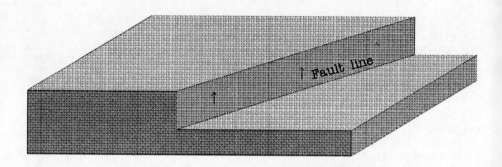

Figure 7.14: A geographical fault

It is relatively easy to perform faulting across the plane produced by the landscape procedure, and the results can be very realistic. After lengthy experimentation I have found that approximately 100 faults produce very realistic landscapes. The program in Listing 7.8 will randomly produce 100 straight faults, after every 10 of which the landscape plane is displayed. After the process is totally complete the resulting landscape is drawn and kept on the screen until a mouse button is pressed.

The method used to generate faults here can best be understood by following the sequence on the screen, as this is where all the processing occurs. First a box is drawn on the screen around the area $x=0$, $y=0$ to $x=47$, $y=47$ – the enclosed area represents the plane. A straight line is then randomly drawn from one side of the box to another. The area on one, randomly selected, side of the line is then filled in using the FILL command. A nested pair of FOR...NEXT loops containing a call to the POINT function are then used to determine the colour of each pixel in the area representing the plane. All those which are filled in then have their corresponding point on the 3D landscape plane raised (or lowered) by a uniformly random amount, the rest of the points being left as they were. Figures 7.15(a) to 7.15(d) show the faulting program in action.

Determining which points lie on which side of the line using the FILL and POINT commands is something of a cheat to avoid mathematics, but it is very effective and also means that a range of differently shaped faults can be accommodated. For example, a lunar surface can be generated by randomly placing randomly sized circles inside the box to create crater effects.

Imitating Nature - Fractal Landscapes

```
@Check_res
Hidem
'
Dim Z%(47,47)
Dim Poly_x%(95)
Dim Poly_y%(95)
Dim Side(1),X(1),Y(1)
'
For Iteration=1 To 100
'
Rem Select sides to draw line between
Side(0)=Random(4)
Repeat
Side(1)=Random(4)
Until Side(0)<>Side(1)  !Make sure that they're different
'
Rem Pick a random position on each side to draw between
For F=0 To 1
If Odd(Side(F))
X(F)=Random(48)
Y(F)=((Side(F)-1)/2)*48
Else
X(F)=(Side(F)/2)*48
Y(F)=Random(48)
Endif
Next F
'
Repeat
Change%=Random(5)-2  !Select random change in height
Until Change%<>0  !for -2 to 2 (but not 0)
"
Rem Plot fault on screen
Cls  !Clear the screen
Print At(1,8);"Iteration:";Iteration
Box -1,-1,48,48  !Draw the plane boundary
Draw X(0),Y(0) To X(1),Y(1)  !Draw the fault line
Deffill 1,2,8  !Set fill pattern to solid black
Fill Random(2)*47,Random(2)*47  !Fill one side of line
'
Rem Alter z% array from screen data
For X=0 To 47
For Y=0 To 47
If Point(X,Y)>0  !If colour of point is black...
Z%(X,Y)=Z%(X,Y)+Change%  !...alter the height by change%
Endif
Next Y
Next X
'
Rem Plot landscape every ten iterations
If Int(Iteration/10)=Iteration/10
Gosub Landscape
Endif
'
Next Iteration
@Waitmouse  !After 100 iterations wait for button
```

Listing 7.8: Program to produce a series of faults and draw the resulting landscape

Figure 7.15a - d: The process after 10, 40, 75 and 100 iterations

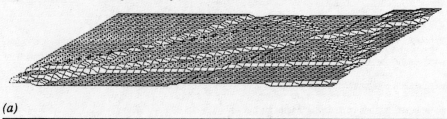

(a)

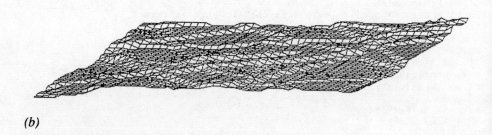

(b)

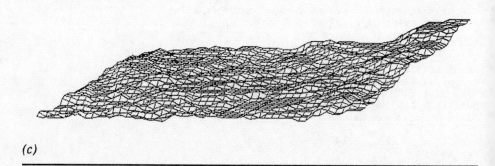

(c)

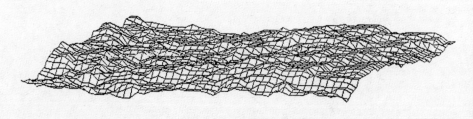

(d)

Using Landscape Techniques to Plot Fractals in 3D

Earlier in the book it was suggested that some fractals, such as the Feigenbaum diagram and Mandelbrot set may benefit from being displayed in three dimensional landscape form. As we now have the landscape procedure, which will produce a landscape from the contents of the two dimensional $z\%$ array, such images are easy to create. The same processes and calculations as those used in the original fractals must be followed of course, but instead of plotting to the screen values are placed in the $z\%$ array and plotted later by calling landscape.

The Feigenbaum Diagram

In Chapter 2, where the Feigenbaum process was discussed, it was mentioned that colour could be used to represent the number of times a point was plotted, but colours did not present this very clearly. A more effective way of showing this information is to draw the Feigenbaum diagram in three dimensions, with the height of each point representing the number of times a point was plotted.

This is very easy to perform using the landscape procedure, as the $z\%$ element for each point can be incremented each time it is visited, eventually giving a height proportional to the number of times the point would have been plotted. The program to draw the 3D diagram is given as Listing 7.9 and the result is shown in Figure 7.16.

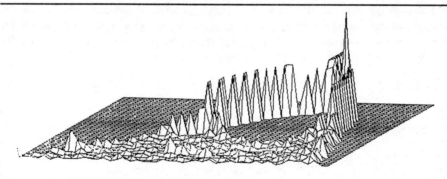

Figure 7.16: Feigenbaum landscape

It is easy to see here that the period one section is the highest part of the diagram (raised 50 pixels above the plane) but at bifurcation the height of the points is halved (to 25). In the chaotic region at the far end of the plane the points vary in height, but are generally very low (one or two pixels high).

```
@Check_res
'
Dim Z%(47,47)
Dim Poly_x%(95)
Dim Poly_y%(95)
'
Print "Calculating..."
'
For C=1.8 To 3 Step 0.025
Y=Int((C-1.8)*40)
P=0.3
For Iteration=0 To 100
If Iteration>25
X=P*34
Inc Z%(X,Y) !Raise point by one pixel
Endif
P=P+C*P*(1-P)
Next Iteration
Next C
'
Gosub Landscape
@Waitmouse
```

Listing 7.9: Program to produce the Feigenbaum diagram in Figure 7.16

Much of this listing is familiar, the only difference being that instead of scaling the co-ordinates up to the size of the screen they are scaled to fit on the 48x48 point plane. The corresponding fact that only 48 values of *c* need to be tested makes the 3D method considerably faster than plotting to the screen. Because of the variety of height on the plane it is quite difficult to recognise the Feigenbaum diagram when it is drawn in this way, as the foreground obstructs most of the diagram. To see the structure of the diagram more easily points on the plane can be lowered in proportion to the number of times they were visited, rather than raised. This can easily be achieved by changing the inc z%(x,y) line to read:

```
Dec Z%(X,Y) !Depress point by one pixel
```

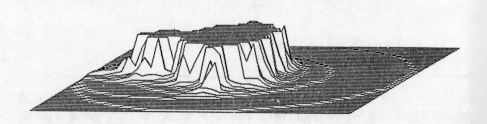

Figure 7.17: Mandelbrot landscape

```
@Check_res
'
Dim Z%(47,47)
Dim Poly_x%(95)
Dim Poly_y%(95)
'
Box 0,0,47,47
'
Rem User editable constants
Amin=-2 !Lowest value of a
Amax=2 !Highest value of a
Bmin=-2 !Lowest value of b
Bmax=2 !Highest value of b
Max_iteration=33 !Iteration ceiling
'
Rem Calculate ranges, offsets and multipliers
A_range=Amax-Amin
B_range=Bmax-Bmin
A_mult=47/A_range
B_mult=47/B_range
A_offset=(0-Amin)*A_mult
B_offset=(0-Bmin)*B_mult
'
For A=Amin To Amax Step (A_range/47)
For B=Bmin To Bmax Step (B_range/47)
P=0
Q=0
Iteration=0
Repeat
Pnew=P*P-Q*Q+A
Qnew=2*P*Q+B
P=Pnew
Q=Qnew
Inc Iteration
Until P*P+Q*Q>=4 Or Iteration=Max_iteration
Xp=Fn Round(A*A_mult+A_offset) !Calculate pixel position
Yp=Fn Round(B*B_mult+B_offset) !
Z%(Xp,Yp)=Iteration !Place height in z% array
If Odd(Iteration)
Plot Xp,Yp !Plot to screen
Endif
Next B
Next A
'
Rem Plot landscape
@Landscape
@Waitmouse
'
Rem The ROUND function is used to round off numbers
Deffn Round(Float)=Sgn(Float)*Int(Abs(Float))+Int(2*Frac(Abs(Float)))
```

Listing 7.10: 3D conversion of the Mandelbrot plotter in Listing 4.4

The Mandelbrot Set

As explained in Chapter 4, the coloured contours around the Mandelbrot set are used to indicate the number of iterations required to free the corresponding point from the circle. The drawback of coloured contours is that the ST's palette of 16 (or 2) colours makes it impossible for each iteration number to have a unique colour. By plotting the set in three dimensions this problem is overcome because each iteration number can have a unique height instead (provided that the iteration ceiling does not exceed the vertical resolution of the screen). Listing 7.10 is a 3D conversion of the multi-purpose Mandelbrot plotter given in Listing 4.4, it will draw any section of the set (specified using the variables at the beginning of the program) in three dimensions. The result of running the program with the default values of `xmin=-2`, `xmax=2`, `ymin=-2`, `ymax=2` and `max_iteration=33` is shown in Figure 7.17.

As in the Feigenbaum diagram, the program is much faster than the equivalent two dimensional one because it is scaled to plot to the 48x48 plane rather than the whole screen. In order to trace the progress of the calculations a tiny replica of the plane is produced while the array is being filled, before the call to `landscape` is made. Unlike the faulting method of landscape generation this replica is not used during the creation of the 3D Mandelbrot set.

As Figure 7.18 shows, the set can be inverted by altering the line which sets the height of each point in the z% array (`Z%(Xp,Yp)=Iteration`) to read:

`Z%(Xp,Yp)=-Iteration`

The method and scaling detailed here can also be used to plot three dimensional Julia sets.

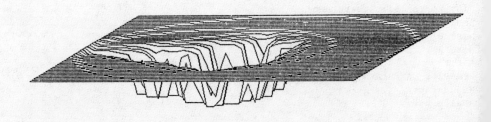

Figure 7.18: Depressed Mandelbrot landscape

Further Ideas for Fractal Landscapes

As indicated earlier in this chapter there are many techniques available for generating natural looking landscapes, but only a small selection have been discussed here. Details of more complicated, and more realistic, landscape generators can be found in the books listed in Chapter 9. Even if you find the naturalness of the landscapes given here acceptable it is still possible to improve their appearance in a variety of ways, described below.

The techniques used to plot the Mandelbrot set and Feigenbaum diagram may also be enhanced to produce better quality output, and both Martin fractals (discussed in the next chapter) and Julia sets may be drawn in landscape form using the methods described above.

Shading

Colour can be put to good use when drawing both pseudo-natural and fractal landscapes. It is particularly suited to landscapes which are composed of tiny quadrilaterals, because each can be filled in with a colour determined by some shading method. The easiest of all colouring methods involves filling areas of water in blue and areas of land with green. However, the simplicity of this solution is such that the black divisions between the segments of landscape still need to be drawn to separate foreground hills from those further back.

Ideally the segments of land should be coloured differently depending on their height, or maybe depending on the angle they make with the vertical and the direction in which they face. This is one of the few occasions when using a colour monitor can actually increase the apparent resolution because each colour segment of land can be as small as one pixel, unlike the monochrome segments which need to be larger to convey their position and inclination properly.

Of course the best (but slowest) shading method would be ray tracing, the intricacies of which are best left to a specialised book on 3D graphics. Such a book would also explain rotation in three dimensions which would allow landscapes to be viewed from different angles, making more of their structure visible. Although experimentation with shading is to be recommended, three dimensional techniques of this complexity often require significant mathematical and programming skill.

Trees and Bushes

A interesting addition to a pseudo-natural landscape image is foliage. The plant generation techniques discussed in the previous chapter can be applied to create a variety of trees and bushes on landscapes. The starting and replacement strings can be made to vary from plant to plant using the same type of inheritance as that discussed above. Each plant could have random description strings generated based on those of neighbouring plants, causing realistic grouping of similar plants. Facts about possible growing positions such as height, gradient and proximity to water could also be used when determining what type of plants the computer should place where. Adding plants can be very enlightening, but a difficult task if approached in the wrong way. However, the following guidelines should make it relatively easy to incorporate plants into a landscape generator:

❑ It is more convenient to place plants so that their trunks emanate only from the points where four segments meet, because the elevation of these points above the plane can quickly be found by interrogating the $z\%$ array.

❑ The plant drawing routine does not need to be altered because the plant must still be drawn with a vertical trunk, and a two dimensional rendering is still perfectly acceptable.

❑ The addition of plants on a monochrome system is not advisable because they tend to get lost among the lines that make up the landscape.

Imitating Nature – Cell Culture

Unlike the two previous chapters on imitating nature this one does not set out to copy a natural object using a set of rules, but rather shows how a purely mathematically based process can generate images which resemble natural structures. The title is actually very much of a generalisation as the images shown here could be said to resemble a range of natural objects, the only similarity being that they look like the type of thing you would expect to see through the microscope of a biology laboratory.

The Martin Process

Of course the common factor linking these images is the process used to draw them, developed by Barry Martin of Birmingham's Aston University. The process is very similar to those covered at the beginning of the book in that a pair of variables are repeatedly transformed by two non-linear equations and the results plotted on a two dimensional plane, as shown in Figure 8.1 (overleaf).

This process is particularly similar to the one used to draw the Lorenz attractor except that only two variables are used, rather than three, and the points are simply plotted rather than being joined together. Also, the Martin equations (shown below) are considerably simpler, relying only on a few elementary functions.

$x_{new} = y - \text{SGN}(x) * \text{SQR}(\text{ABS}(b*x-c))$

$y_{new} = a - x$

Here x and y are the two variables and a, b and c are constants similar to those found in the Lorenz equations. The important parts of these equations are the SGN, ABS and SQR functions, whose GFA BASIC interpretations are shown below.

Sgn(x) is used to determine the sign of a number (whether it is positive or negative). The function returns 1 if $x>0$ (positive), -1 if $x<0$ (negative) and 0 if $x=0$.

Sqr(*x*) returns the square root of a number ($\sqrt{x}$). So, for example, Sqr(9) would return 3, because 3*3 = 9. More information on square roots and indices can be found in Appendix B.

Abs(*x*) returns the absolute value of *x*, in other words the distance of *x* from 0. This is sometimes useful for checking if a variable is within a certain range. For example if abs(x) <=3 could be used to test whether *x* is between –3 and +3 inclusive.

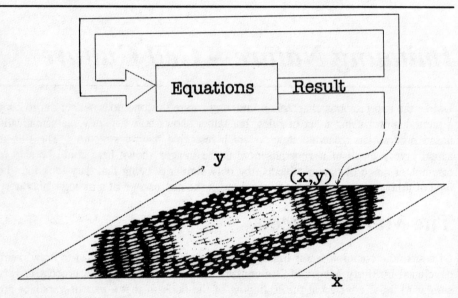

Figure 8.1: Relationship between the two equations and a Martin fractal

The combination of these functions and the feedback between *x* and *y* (shown in Figure 8.2) on successive applications of the equations makes the Martin process unpredictable and immensely complex. By plotting *x* against *y* on the ST's screen, after each iteration, two dimensional Martin fractals can be created. These demonstrate just how chaotic and varied the output from these two simple equations can be.

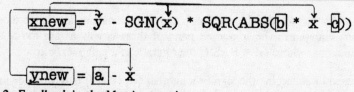

Figure 8.2: Feedback in the Martin equations

A program to perform the Martin process and simultaneously plot the resulting fractal is shown in Listing 8.1. Due to the large variety of output that can be generated with this program several facilities have been included specifically to aid experimentation.

```
@Check_res
'
Rem Request constants
Input "a=",A
Input "b=",B
Input "c=",C
Cls
'
Rem Set initial values for variables
X=0
Y=0
I=0
'
Hidem
'
Rem Nested loops to draw fractal and check for mouse button
Repeat
Repeat
'
Rem Perform Martin process
Xnew=Y-Sgn(X)*Sqr(Abs(B*X-C))
Ynew=A-X
X=Xnew
Y=Ynew
'
@Fplot(150+X*0.4,100-Y*0.4) !Plot point
Inc I !Increment iteration count
Until Mousek>0 !If mouse button is pressed...
Message$="Iterations:"+Str$(I)+"| |Do you want to leave|the program?"
Alert 1,Message$,2,"Yes|No",Button !Display alert box
Until Button=1 !If user selects YES...
Edit !return to editing screen
```

Listing 8.1: Program to perform the Martin process and plot the result

The main feature of this type is an alert box, drawn on the screen in response to a mouse button being pressed during program execution. As well as displaying the number of iterations completed the box also provides the opportunity of exiting the program. This ensures that the chances of accidentally leaving the program are reduced and compilation is more feasible because it is not necessary to test the keyboard for the *stop program* key combination (SHIFT-ALTERNATE-CONTROL).

Unfortunately, keeping track of the iteration count slows computation down slightly, so once you are better acquainted with the program you may wish to remove all references to the *i* variable, as it is this which stores the iteration count.

Another feature which makes the program suitable for immediate compilation is the inclusion of INPUT commands to request the constants from the user, eliminating the need to delve into the program code to change the nature of the fractal produced.

Upon running the program you will be prompted to enter the three constants. The values entered for these determine which of the many possible fractals will be drawn. To pilot the program and to demonstrate a typical Martin, if there is such a thing, the following constants can be used:

```
a=45
b=2
c=-300
```

After entering the above values, in response to the relevant prompts, you will soon (during the first 5,000 iterations) see a small group of near-perfect circles appear, connected together by pairs of lines, as shown in Figure 8.3(a).

At this stage it looks more like a network of underground passages than a biological construction. After a short wait (26,000 iterations complete) all of the exits from this network are blocked off and a collection of tiny lines appears in the space outside, see Figure 8.3(b) (overleaf).

Figure 8.3(a): Martin fractal where a=45, b=2, c=-300 after 5,000 iterations

Following another, very brief, pause (28,000 iterations complete) there is a sudden 'burst' of growth moving out from the central passages. This growth is much more sustained than before, continuing until around the 100,000th iteration. The scene after this burst is shown in Figure 8.3(c) (overleaf) and clearly highlights the new, more natural looking, growth dwarfing the original network.

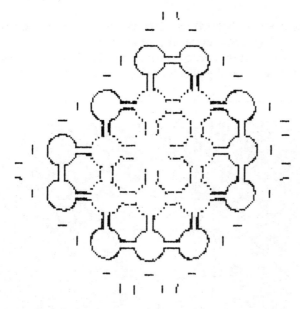

Figure 8.3(b): The Martin fractal after 26,000 iterations

Figure 8.3(c): The Martin fractal after 70,000 iterations

Further bursts occur at the 400,000 and 600,000 iteration marks which thicken up existing lines and fill in gaps, eventually producing the picture shown in Figure 8.3(d). The main features of this fractal, such as the high level of complexity and the bursting phenomenon are typical products of this program.

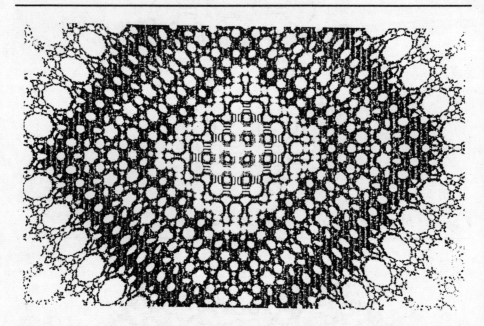

Figure 8.3(d): The Martin fractal after 1,000,000 iterations

Like many other fractals it is very interesting to watch this one grow. By slowing the program down it can be seen that the image is built by pixels being dropped, as if from an invisible satellite in irregular orbit around the centre of the screen. Using this analogy the periods of inactivity seen between the bursts in the above example can be said to be caused by the satellite dropping pixels on top of ones already on the screen, thus making no visible change. This infers that the satellite has predictable periodic behaviour, but this can never be the case with a non-linear dynamic system.

In fact, during periods of inactivity, the satellite is following a subtly different path, too similar to be noticed in the crude approximation produced on the screen. However, the same type of positive feedback that causes the butterfly effect amplifies this small deviation and eventually it suddenly becomes very large, causing the satellite to assume a much different path, noticeable on the screen as a sudden burst of new pixels.

Return of the Butterfly

The notion of non-periodicy and the butterfly effect in a system such as this will already be well known from Chapter 3, but the bursting phenomenon is far more visual here than in the Lorenz attractor. These kinds of erratic flurries of activity are also familiar to share dealers, stock brokers, weather forecasters and other people whose professions are concerned with the world of non-linear dynamics. In these areas of the real world such bursts manifest themselves as sharp swings in share prices, sudden drops in temperature and rapid increases in oil prices. It is highly probable that such sudden unexpected events are created by previously unnoticed small changes getting larger by ever increasing amounts.

The now legendary stock market crash of 1987 was apparently due to dealers over-reacting to a small devaluing of stocks, sending shares into a downward spiral. The fall was further magnified by the newly installed computer dealing systems which sold shares instantaneously, without the protective natural damping provided by slower human reactions. Nobody could have noticed the start of this spiral in advance because it would have been too small, meaning that such events can only be spotted when their magnitude is such that they are hard to prevent.

A similar manifestation of the butterfly effect could cause a system, apparently continuing normally, to suddenly collapse without warning. This is an interesting suggestion and has precipitated much debate in recent years. The fact that the weather and the Earth's eco-system are vulnerable to this type of positive feedback makes the possibility of a sudden end to the human race quite feasible.

Further Experimentation with the Martin Program

The number of different images that can be produced with Listing 8.1 is truly daunting, so a selection of some of the most interesting constants are shown in Table 8.1. Note that although most of these examples will fit on the screen using the current scaling, some of them may benefit from being enlarged. This can easily be done by changing the co-ordinate multipliers in the `fplot` line. For example the following line will produce output of twice the size:

```
@Fplot(150+X*0.8,100-Y*0.8)
```

The position of the image on the screen may also need to be altered. This can be done by editing the two large numbers in this line, as it is these which collectively determine the offset. An example of a Martin enlargement is shown in Figure 8.4.

Table 8.1: Some interesting constants for the Martin program

a	b	c
68	75	83
90	30	10
10	-10	100
-200	-4	-80
-137	17	-4
10	100	-10
-137	17	4

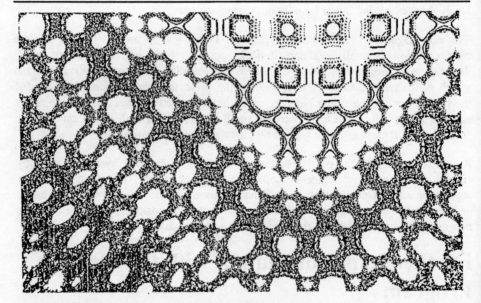

Figure 8.4: Mathematical enlargement of Figure 8.3(d)

Extensions to the Martin Program

There is little point in adding an expeditious user interface to the Martin program because very little input is required. However, there are a few simple enhancements that can be included in other areas of the program:

Colour

If you have a colour monitor you may want to add colour to the Martin fractals in order to gain a fuller understanding of the process which creates them. To do this there must be something that the colours can represent, as discussed in earlier chapters. Here they can be used to show the number of times a particular pixel position has been visited. As well as making the fractals more aesthetically pleasing this also makes it possible to see which points are being plotted at any given time, even during a period of inactivity.

The simplest, but probably slowest, way to facilitate colour is to work out the position of the next pixel to be plotted, get the colour of the screen at this position using the POINT function, put it in a variable, increment it and then plot the pixel in the colour denoted by the variable. By doing this the colour of a point on the screen will change every time a pixel is plotted on it. Using this method has the advantage that everything is stored on the screen, so no extra variables or arrays need to be declared.

Exactly the same technique was used in Chapter 2 to add colour to the Feigenbaum diagram. It can by implemented in the Martin program by replacing the fplot line with the program fragment shown below. Note that, because of the 16 ink limit, the variable containing the colour value is of modulus 16, so after a point has been visited 16 times it returns to being white and the colour cycle starts again.

```
Xp=150+X*0.4 !Calculate x and...
Yp=100-Y*0.4 !...y position of point
Oldcolour=Point(Xp*Monitor,Yp*Monitor) !Determine old colour
Color (Oldcolour+1) Mod 16 !Set new colour
@Fplot(Xp,Yp) !Plot the point
```

Sound

If you have recovered from the musical Feigenbaum diagram in Chapter 2 you may want to add sound to the Martin program. This can be very useful for alerting you to pixels being plotted on the screen which would otherwise have gone unnoticed. The most efficient way to create such a *pixel alarm* is to start a tone whenever a new pixel is plotted, and halt the tone when one is plotted on an already occupied space. Again, the POINT function can be used to check the intended plotting position to see whether it is free. To add the alarm the routine below should be included in the program given in Listing 8.1, in place of the fplot line. Note that, in an effort to increase program speed, no pixel is plotted if the chosen space is already full.

```
Xp=150+X*0.4 !Calculate x and...
Yp=100-Y*0.4 !...y positions
If Point(Xp*Monitor,Yp*Monitor)=0 !If position is empty...
Sound 1,15,1,4,0 !...start tone and...
@Fplot(Xp,Yp) !...plot point
Else
Sound 1,0,1,7,0 !Otherwise stop tone
Endif
```

This technique allows even the smallest of additions to the fractal to be noticed, and bursts of activity can easily be distinguished due to their characteristically continuous tones. This relieves you of the boredom of having to stare at the screen in order to be sure of seeing all of the important changes. Making the pitch of sound produced proportional to the x or y variable may be a way of further enhancing the program.

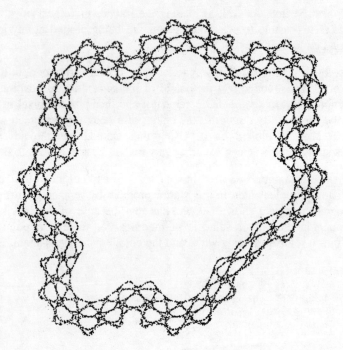

Figure 8.5: Martin fractal with a=-137, b=17, c=-4 after 30,000 iterations

Compilation

The easiest way to improve the program, if you possess a GFA BASIC compiler, is to compile it. The Martin program is ideal for this as it is totally self contained and most variations of the fractal can be generated without the need to change any of the program code. If you do compile it do not forget to choose the *STOP NEVER* option, as there is no need to test for the SHIFT-ALTERNATE-CONTROL key combination.

Further Experiments with Natural Fractals

This concludes our experiments with the imitation of nature, but there are obviously many natural objects that we have not attempted to model, including clouds, smoke, sparks and animals. Some of these can be generated using techniques already discussed in this book, but others will require the formulation of some kind of specialised algorithm. Such algorithms can either be constructed by examining real life objects, as in the last two chapters, or they can be gleaned from other chaos books. Either way the artificial creation of natural images is a very interesting application of computer science, and one which is bound to become very popular as advances in graphics facilities and processing power are made in the future.

Imitating Nature - Cell Culture 155

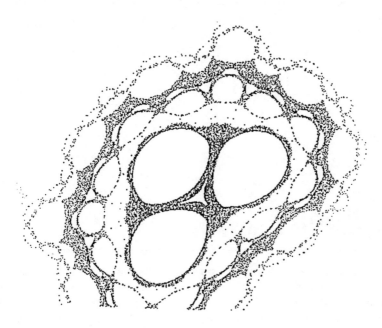

Figure 8.6: Martin fractal with a=68, b=75, c=83 after 40,000 iterations

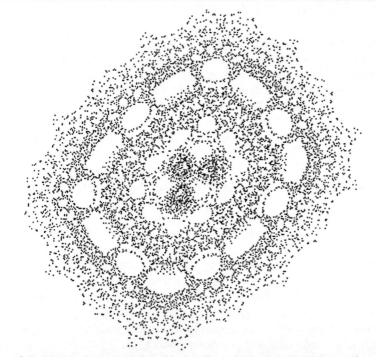

Figure 8.7: Martin fractal with a=12, b=18, c=256 after 10,000 iterations

The Future

Although the initial chaos boom has now subsided, the outlook for applied chaos research looks promising. There are now many dedicated chaos researchers and it is very probable that the number of prominent scientists interested in chaos will grow dramatically in the future as the older generation of scientists is replaced by one more willing to use computers.

Can the Future of Chaos be Predicted?

In any science there is normally a time lag between leading edge research and the state of the amateur scientist's contribution, due to professional researchers having superior resources. This is true in chaos, and by browsing through recent publications such as those listed in the bibliography it is possible to predict the nature of future home computer based research. Because the advances in computer technology are so fast the chaos time lag is shorter than in most sciences, at around 10 years. To provide an insight into the home computer chaos research of tomorrow, a few of the more interesting areas of today's professional activity are discussed below.

Fractal Maths in Data Compression Applications

The advantages of data compression are obvious, money can be saved in telephone bills when sending files via modem and for any given file less storage space is required on a computer's storage device. After discovering that something as complex as the Mandelbrot set can be described by one simple equation (see Figure 9.1(a)) scientists speculated that all images may be able to be treated like fractals and compressed so as to be described by a small equation, as shown in Figure 9.1(b). Theoretically such compression is possible, but with current technology the complex fractal mathematics involved in compressing a single television quality image in software takes such a long time that it is of no practical use.

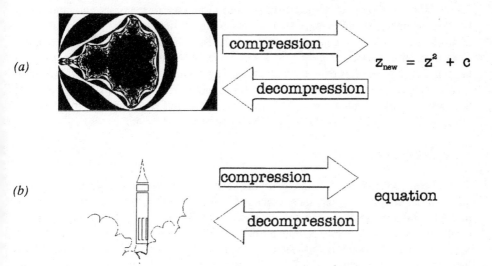

Figure 9.1 (a) Compression and decompression of the Mandelbrot Set to and from its equation. (b): Mathematical compression and decompression of a photograph

Recently, however, an American team co-led by prominent chaos writer Michael Barnsley claims to have developed an IBM PC-based image compression system which can compress and decompress moving VGA graphics (640x480 pixels, 16 colours) in real time. The system can run in real time because most of the processing is carried out by a purpose-built expansion card, which has its own high speed microprocessor. The incredible speed is matched by the reduction in data size, which allows almost one minute of moving video images to be squeezed onto one 800k floppy disk. The extent to which the system relies on state-of-the-art hardware is reflected in the price of $25,000, which obviously puts it out of the reach of most PC users. Since a computer image is constructed from digital data it should be possible to extend this technique to other forms of computer data, including text and program code.

Performing such data manipulation on the ST would be almost impossible, considering that the decompression of the Mandelbrot set, from its equation, can take several hours (depending on the resolution and language used).

Telephones with Minds of their Own

The vast telecommunications network that now stretches across the world has also been attracting the attention of chaos researchers. It has been suggested that tiny bursts of static interference in telephone lines can be amplified by positive feedback at telephone exchanges and other installations causing them to grow into large, audible sounds. Arthur C. Clarke first proposed this idea many years ago, and it has now been embraced by science-fiction fans who can imagine the world telephone network acting like a complex biological organism, dialling up people and sending them intelligible messages and faxes at will. In reality feedback on this scale does not happen but stories of people's telephones doing strange things, normally explained with reference to ghosts and the underworld, could well be caused by this typical manifestation of the butterfly effect. This type of chaos is not unique to telephone lines though, it could also occur in video and computer networks.

The New Art

The recent publication of a range of books dedicated to fractal generated plants, animals, landscapes and abstract patterns has helped to establish fractal graphics as an important art form in its own right. This is one area of current interest which can be explored with humble home computers like the ST, as described in Chapters 6, 7 and 8. Using the picture saving and loading routines given in Appendix A it is possible to load fractals created using GFA BASIC programs into Degas and create a wide range of original compositions, an example of which is shown in Figure 9.2.

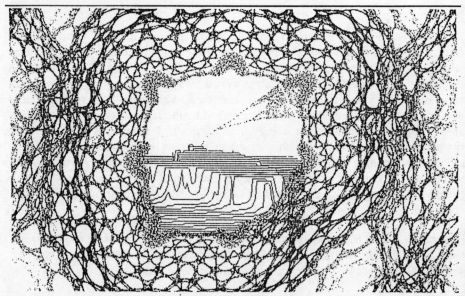

Figure 9.2: Original fractal composition produced using Degas and programs from this book

Chaos can also be applied to one of the other popular programming areas, ray tracing. If you have ever experimented with ray tracing shading algorithms you will know that creating the illusion of a naturally rough surface, such as a carpet or road, is no easy task. An interesting technique for overcoming the problems normally associated with such surfaces is to use fractal landscape generation techniques, shown in Chapter 7, to create fairly flat landscapes on surfaces designated as being rough. The light incident on such surfaces will then appear to be reflected naturally. Such sophisticated ray tracing on the ST would be very time consuming, but it is not impossible.

What Use is Chaos?

As discussed in Chapter 4, the computer press have been keen to demonstrate the use of chaos for drawing pictures but are reticent about the relevant theory. Even when the implications are discussed the attitude adopted is generally negative, with emphasis being placed on chaos hampering weather predictions and possibly being associated with the end of civilisation. However, chaos is like fire, it is unpredictable and destructive. Yet if its power is tapped it can be used in a variety of useful applications. Although fractal maths has been used in image compression, few applications have yet been written which actually use chaos. An example of how powerful a tool chaos could be is clearly visible in nature, in the anti-viral immune system of the human body.

When a virus enters the body the immune system produces a random variety of anti-bodies and sends them into the blood stream. The progress of each type of anti-body is then monitored and after a period of time the identity of the successful one is fed back to the immune system. The body then produces the thousands of clones of the successful anti-body which are necessary to kill off the virus. This use of randomness and feedback (i.e. chaos) provides the body with a system which can cope with an enormous range of viruses, even new strains to which the body is not accustomed. This method has significant advantages over the equivalent deterministic process, which would need to know the details of every virus that currently exists and all those which will exist in the future.

The evolutionary process of natural selection also uses chaos, to preserve life on earth. New organisms are produced randomly through mutations and the success of each organism is fed back into the system in line with the 'survival of the fittest' theory. By moving forward on a broad, ever changing front nature can ensure that life still exists even in the bleakest of conditions, when the dinosaurs became extinct for example.

Derivation of Pi Using the Monte Carlo Method

The power of controlled randomness can also be demonstrated by considering the challenge of finding the value for the commonly used mathematical constant, pi (Π). Interrogating a pocket calculator, or GFA BASIC, will give a pre-determined value for pi in the region of 3.141592654 but pi is irrational, meaning that an infinite number of decimal places are required to convey its value accurately. Many people have experimented with pi over the centuries and many different ways of deriving its value have been devised, the most accurate of which recently yielded an answer with 10 billion decimal places. Not surprisingly, most of these methods are deterministic but one, known as the Monte Carlo method, can be based on something as random as the scattering of raindrops on the ground.

Before executing this method it is necessary to find a large area of unsheltered flat land, on which the pattern shown in Figure 9.3 should be drawn as accurately as possible. Before describing the next stages of the process some peculiarities of the pattern must be considered. As the figure shows, it consists of a circle, of radius two metres, inside a square of side four metres. Remembering that the area of a rectangle is its height multiplied by its length and that for a circle it is pi multiplied by the square of the radius:

Area of square   =   4*4   =   16
Area of circle   =   PI*(2)^2   =   4*PI

The ratio of the area of the circle to the area of square is therefore:

$$4*PI:16 \ = \ \frac{4*PI}{16} \ = \ \frac{PI}{4} \ \text{or} \ PI/4$$

Note that the actual size of the pattern on the ground in not important, as long as the ratio of the area of the circle to the square is constant at pi/4. However, the larger the area the more accurate the results will be for reasons discussed later.

After inscribing the pattern on the ground it is necessary to wait for a light rain shower to occur. After the shower the total number of raindrops which landed in the entire square area (the number of drops captured) should be recorded. Of these, the number which fell inside the circle (the number of *hits*) should also be noted. The hit ratio, the ratio of hits to raindrops captured, should then be calculated. Because the raindrops can be assumed to have fallen totally randomly we can say that they were distributed evenly across the square area. Since the ratio of the area of the circle to the area of the square was calculated above as pi/4 it follows that the ratio of the number of hits to the number of raindrops captured also equals pi/4. Therefore, by multiplying the hit ratio by four, pi can be determined.

In practice this method would prove very time consuming, as a large area would be needed to determine pi with any accuracy meaning that a large number of raindrops would need to be counted. However, the outcome of the process can be speeded up by modelling it in GFA BASIC, this also eliminates the problems associated with waiting for a light shower and ensuring that the raindrops do not evaporate before the counting is complete.

The Future

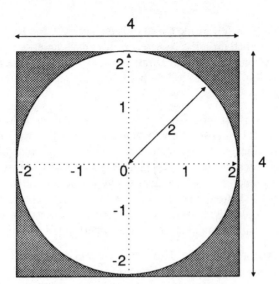

Figure 9.3: Pattern used when determining the value of pi

A program to simulate the process and calculate the result after each drop has landed is shown in Listing 9.1. This uses a pattern of the same size as that shown in Figure 9.3. This size is convenient because the circle is the same as the one used to produce the Mandelbrot set, so exactly the same equation can be used to determine whether a given drop fell inside or outside the circle.

```
Rain_drops%=0 !Set initial rain-drop count...
Hits%=0 !...and hit count to zero
'
Do
'
X=Rnd*4-2 !Select random x position
Y=Rnd*4-2 !Select random y position
'
Inc Rain_drops% !Increment the rain-drop counter
'
If X^2+Y^2<4 !If drop lies inside circle...
Inc Hits% !...increment hit counter
Endif
'
Result=4*Hits%/Rain_drops% !Calculate 4*ratio (i.e. PI)
Print At(1,1);"Rain-drops:",Rain_drops%
Print "PI:",Result !Print result
Loop !Repeat indefinitely
```

Listing 9.1: The raindrop program for determining pi

After initialisation of the raindrop and hit counts, the program enters a continuous cycle in which a point on the board is selected at random within the range $-2<x<2$ and $-2<y<2$ and the raindrop count (held in `rain_drop%`) is incremented. The point is then tested and, if it fell inside the circle ($x^2+y^2<4$), the `hits%` variable is also incremented. After each drop has landed (i.e. after each iteration) the ratio of `hit%` to `rain_drop%` is calculated and multiplied by 4 to give the value of pi, which is printed on the screen. After several thousand iterations the first few digits of pi have been determined.

This is one of the slowest processes for deriving pi but it does show how randomness can be put to a useful purpose. If civilisation knew no other way of calculating pi this method would obviously be very important. The model produced by GFA BASIC is flawed, however, because the random number generator is not truly random causing accuracy to be limited. Also the random nature of the algorithm causes the result to drift annoyingly away from the real value of pi during some stages of processing, although the laws of statistics lead us to believe that the accuracy of the result does generally increase with the number of iterations performed.

The way that the ratio drifts randomly around pi/4 is shown in Figure 9.4, which clearly demonstrates that pi/4 is the attractor of the process. The graph is similar to the deterministic time series shown in Chapter 2 (Figure 2.7(a)), where the population was attracted to 100%, but the oscillations around the pi/4 attractor are much less ordered due to their random origins.

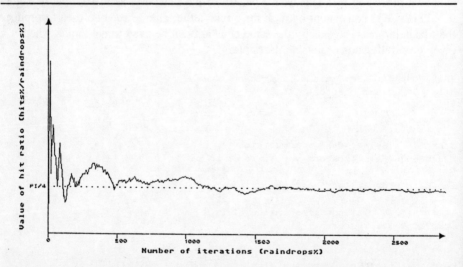

Figure 9.4: Graph showing the hit ratio (drawn as a solid line) drifting around the real value of pi/4 (dotted line)

Chaos techniques such as those used in the immune system would be of particular use in artificial intelligence applications where, at present, all possible scenarios must be considered when developing a set of rules to solve a certain problem. It is even conceivable that future high level computer languages might have built-in chaos commands.

Life After *Computers and Chaos*

I have covered most of the major facets of chaos theory in this book, but the constantly expanding nature of the science means that there are still many more experiments which may be performed on the ST. Even during the closing stages of this book's production new theories and ideas that could have been included were being brought to my attention. A large number of further experiments can be based on those in this book, or on the purpose-written support disk (see front of book). Some ideas for such experiments are discussed in the section below.

Two Headed Mandelbrots and Similar Feigenbaums

There is an interesting piece of information that can be found about the Feigenbaum diagram which forms a good basis for a challenging programming problem. By measuring the length of each section of the diagram, results similar to those given in Table 9.1 may be obtained.

Table 9.1: Length of different sections of the Feigenbaum diagram

Period	Length/mm	Ratio
1	690	
		4.6
2	149	
		4.8
4	31	
		4.4
8	7	

From these it is possible to work out the ratio between the length of one section and the next. Note that the size of the diagram being measured is irrelevant but the bigger the diagram the more accurate the results. If the measuring stage of the experiment is performed accurately enough all the ratios should be of exactly the same value, approximately 4.7 (the mean average of those in the table is 4.6). Mitchell Feigenbaum was the first to discover this common ratio, he managed to calculate it as being *approximately* 4.6692016609 10299097.

This value is only approximate because, like the pi constant (Π) discussed above, it is an irrational number, i.e. it has an infinite number of decimal places. To save the trouble of writing out all the decimal places in full mathematicians have labelled it with the Greek letter delta (δ).

Even though the results in Table 9.1 were taken from an enormous print-out of the Feigenbaum diagram the value of delta is still quite inaccurate. A more sensible way of calculating the ratios is by using a BASIC program to determine the exact value of c associated with each bifurcation, and then use these values to calculate delta. A program of this sort would take a long time to execute, but would make a very worthwhile overnight project.

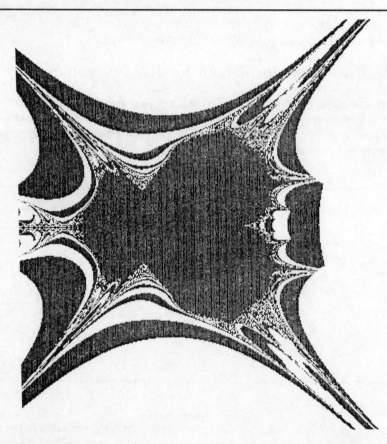

Figure 9.5: 'Spiky' Mandelbrot set

After several years of home computer activity related to the Mandelbrot set it could be assumed that there is nothing further to explore. However, even with this veteran fractal there are many more experiments to conduct. Some variations on the set, such as those shown in Figures 9.5 and 9.6, can be generated by altering the Mandelbrot programs without any preconceived plan. In general, however, it is better to have at least some idea of what effect alterations are likely to have. Some suggestions for simple variations on the Mandelbrot set are described below.

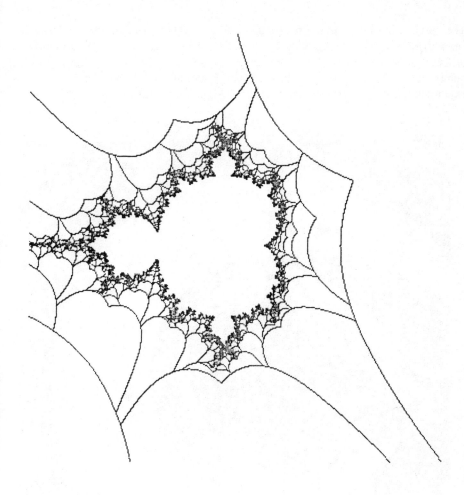

Figure 9.6: Mandelbrot 'web' (based on Listing 4.5)

The first variation on the original Mandelbrot set is caused by altering the lines in the Mandelbrot program which set the initial values of p and q, so that these variables are something other than zero. The changing of these initial values is similar to the Julia process described in Chapter five, but in that process a and b are treated as constants while the initial values of p and q vary according to the position of the pixel being plotted. In the altered Mandelbrot process, however, a and b vary as normal and the initial values of p and q are constants as normal. The result of such a subtle change to p and q can be achieved by replacing the original variable assignment lines in a Mandelbrot program, such as Listing 4.4 with:

```
p=0.6
q=0.2
```

The output from the resulting program (see Figure 9.7) looks like a cross between a Julia and a Mandelbrot set, reflecting the hybrid process. There are a vast number of different sets that can be created using this method, due to the many possible combinations of *p* and *q*. However, an important thing to remember when experimenting in this way is that both variables should always be in the range of –2 to 2.

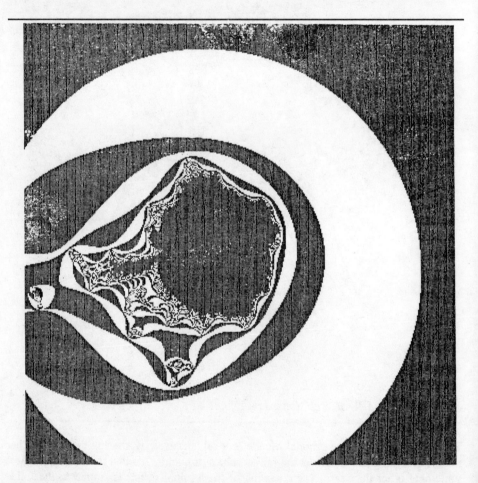

Figure 9.7: Mandelbrot set with amin=-2, amax=2, bmin=-2, bmax=2 and initial values for p and q of 0.6 and 0.2 respectively

Another variation on the set can be achieved by altering the Mandelbrot equation which, recalling Chapter 4, was as follows for the complex number method:

$z_{new} = z^\wedge 2 + c$

This could be expanded to:

$p_{new} = p^2 - q^2 + a$

$q_{new} = 2*p*q + b$

The second Mandelbrot set variation is produced by changing the index of z from 2 to 3. The new equation, with its resulting expansion, is as follows:

$z_{new} = z^3 + c$

$p_{new} = p^3 - 3*p*q^2 + a$

$q_{new} = 3*p^2*q - q^3 + b$

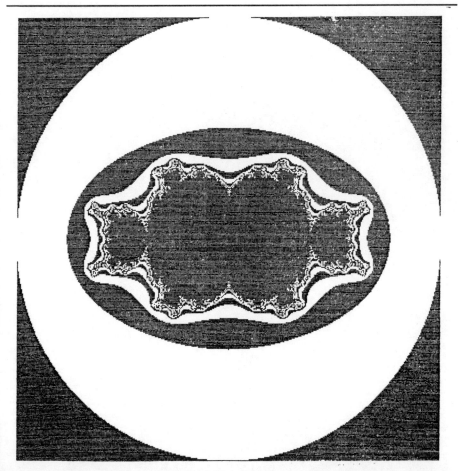

Figure 9.8: 'Two-headed' Mandelbrot set with amin=-2, amax=2, bmin=-2, bmax=2

The two equations for p_{new} and q_{new} can easily be incorporated into a Mandelbrot program such as the one given in Listing 4.4, and will produce output similar to that shown in Figure 9.8. Although the more complicated calculations slow the program down this is still an area worthy of experimentation, try raising z to the power of 4, 5 or even a floating point number (mathematical skills are required to achieve this).

The increased popularity of chaos in the last few years has resulted in many new works becoming available, which collectively describe all the major aspects of chaos theory. If you are interested in creating fractals which have not been covered in this book, or if you want to find out more about chaos history, you will find additional information in the bibliography that follows. It contains brief descriptions of the most notable titles, suggested as a logical progression from this book.

Appendix A

Useful Routines

Throughout this book the example programs have been kept as short as possible to save room and to keep to a minimum the amount of typing, in an attempt to make the theory and principles as clear as possible. This has been done partly by omitting sophisticated user interfaces, and also by gathering together the most widely used procedure definitions and placing them in this appendix, thus creating a central library of general purpose functions to complement those supplied with GFA BASIC. The routines shown here are not specific to drawing fractals, and you may well want to apply them to other applications.

Using the GFA BASIC Procedure Library

As already mentioned, many of the programs lack vital routines, so merely typing the programs in *verbatim* will usually result in a *Procedure not found* error. There are two ways to include the relevant routines. If you are a keen typist you can find out which ones are needed (using the convention described below) and type them in at the end of the relevant programs. Alternatively you can type all the routines in one go, save them as an ASCII file (by pressing <SHIFT> and <F2>), and then MERGE this file (by pressing <F2>) onto the end of any programs that need them. The latter method is by far the most efficient as the procedures only need to be entered once for the whole book. The support disk (see front of book) offers a third alternative, as it provides all the listings from the book with the relevant procedures included in each one. Note that, irrespective of the method used, the procedure definitions must be placed at the end of the program as GFA BASIC halts execution when it reaches these.

A simple convention has been used in the example programs throughout the book to distinguish between calls to general purpose procedures, such as those detailed here, and program specific procedures. If the procedure is called using GOSUB then it is specific to the program in which the call appears, and can be found in or near that program. However, if the procedure name is called using the @ character (the GFA BASIC abbreviation for GOSUB) it is a general purpose routine, and the relevant procedure definitions should be included from this appendix. For example:

```
@Fplot(100,100) is a call to a library routine
Gosub Draw_branch(160,180) calls a program specific routine
```

The GFA BASIC procedure definitions are given below, including a short description of how each one works and a discussion of why it is needed. Note that `check_res`, `fplot`, `fdraw` and `waitmouse` are the only routines essential to the example programs. The picture printing and filing procedures are provided for further experimentation.

Procedure Definitions for Example Programs

Check_res

This procedure is included at the beginning of all programs in this book in order to make them compatible with both high-resolution monochrome and low-resolution colour monitors. When called the procedure sets the `monitor` variable (used by `fplot` and `fdraw`) to 1 for low-resolution mode, or 2 for high resolution. Medium resolution is not supported and the user is told this in an alert box if `check_res` is executed in that mode. The resolution is determined using the XBIOS getres() function (opcode=4). When accessed by GFA BASIC using `res%=Xbios(4)` it returns the screen resolution in the variable `res%`, giving 0 for low, 1 for medium and 2 for high resolution.

```
Procedure Check_res
Res%=Xbios(4)
If Res%=1 !Medium resolution not supported
Alert 3,"This program only|works in high or|low resolution modes.",1,"
                                                    Abort ",Dummy
Edit !Quit back to editor (leave program if compiled)
Else
Monitor=Res%/2+1 !1=Low-res, 2=High-res
Endif
Return
```

Fplot

Fplot performs exactly the same action as the GFA BASIC PLOT function, but is sensitive to the `monitor` variable used to make the programs suitable for both high and low resolution graphics modes. Note that the parameters passed should be in the range normally associated with low-resolution mode, i.e. 0<=x<=319 0<=y<=199. If the `monitor` variable is set to 2 for high resolution x and y will automatically be doubled, to make use of the full screen area. A further improvement of this procedure over PLOT is that `fplot` rounds off floating point co-ordinates which are passed to it, instead of just taking the INT value. This latter measure is necessary because the scaling of fractals often yields floating point co-ordinates.

```
Procedure Fplot(X,Y)
Xp=X*Monitor
Yp=Y*Monitor
Plot Int(Xp)+Int(Frac(Xp)*2),Int(Yp)+Int(Frac(Yp)*2)
Return
```

Fdraw

Fdraw draws a line from the current drawing position to the point whose co-ordinates are passed in the procedure call (similar to the standard DRAW TO function). Again, the advantage of this routine is that it is compatible with both types of monitor and correctly rounds off the co-ordinate passed to it.

```
Procedure Fdraw(X,Y)
Xp=X*Monitor
Yp=Y*Monitor
Draw To Int(Xp)+Int(Frac(Xp)*2),Int(Yp)+Int(Frac(Yp)*2)
Return
```

Waitmouse

This procedure simply executes a continuous loop until a mouse button is pressed. This is used at the end of many of the programs given in this book as it allows the output to be viewed, once the program has finished, without the *end of program* alert box getting in the way.

```
Procedure Waitmouse
Repeat
Until Mousek>0
Return
```

Screen Dumps

Because of the large variety of ST compatible printers it would be difficult to include screen dump routines for all of them. Instead, a set of routines is given which allow any Degas printer driver to be loaded into memory and used from GFA BASIC. The advantage of Degas printer drivers is that there are many available in the public domain and there is at least one for every popular printer. Because Degas stipulates that they must be written in machine code, such drivers are also very fast and compact (exactly 2,000 bytes in length).

The criteria for a Degas printer driver is clearly stipulated in the package's manual, and using this specification it is easy to determine how the driver is loaded and what information it needs. The criteria are as follows:

The driver should be a relocatable machine code program, exactly 2,000 bytes in length, with the extension .PRT. It must be given the following pieces of information:

Command:	0=initialise, 1=print
Resolution:	0=low, 1=medium, 2=high
Screen adr:	Where screen memory starts
Palette adr:	Address of a block containing the colour palette
Workspace adr:	Address of a 1,280-byte workspace area

This information should be transmitted to the routine through arguments passed in the following C-type call:

```
print(command,res,screen_address,palette_address,workspace)
```

GFA BASIC provides the C: function for calling such routines. In this case it would be used as follows (where `routine` holds the RAM address of the printer driver):

```
C:routine(command,res,screen_address,palette_address,workspace)
```

Obviously the printer driver cannot be executed until it has been loaded into memory. This can easily be performed using the BLOAD command, but the problem is finding a place to store it. As the ST has such a large amount of memory, the operating system has to keep track of data and programs in it so that they do not overlap or become lost.

For this reason the operating system must be informed of all memory operations. For instance, in order to load the printer routine into RAM it is necessary for the operating system to allocate a memory block using the Malloc() function (opcode=72). In GFA BASIC this function is called using `address=GEMDOS(72,L:size)` where `address` is the memory location of the first byte in the block and `size` is the size of the block in bytes. If sufficient memory cannot be located, zero is returned in the `address` variable.

The `memory_initialise` routine given later allocates three memory blocks, one for the printer routine, one for its workspace and one for the relevant palette data. At the end of the procedure the printer driver is loaded into the 2,000-byte memory area allocated to it. If sufficient memory cannot be found an error message is returned. Note that `memory_initialise` always searches for the printer driver in the root directory of the disk in drive A, if it cannot find it the program will be terminated. You should therefore ensure that the relevant driver for your printer is present in this directory before `memory_initialise` is called.

Once allocated, a memory block stays allocated, even if the block is empty and even when the program that allocated it ends. This means that if the program is run several times in a row the RAM will soon become cluttered. To avoid this all memory must be properly freed once it has been finished with. This is done using the operating system's Mfree() function (opcode=73). A line such as `status=GEMDOS(73,address)` is used to free the block of memory starting at the address stored in the `address` variable. The `memory_free` procedure given below frees all the memory allocated by `memory_initialise`, and should be used to return all the memory to the system once the screen dump routine is no longer needed.

Useful Routines

The actual screen dump action is performed by the `screen_dump` procedure. This calls the routine twice using the C: function, first to initialise it (command=0) and then to execute it (command=1). The mouse pointer is hidden during printing so that it does not come out on the screen dump. `Screen_dump` loads the palette information into the allocated memory block using the `store_palette` procedure (also given below).

```
Procedure Memory_initialise
Routine=Gemdos(72,L:2000) !Reserve space for screen dump routine
Workspace=Gemdos(72,L:1280) !Reserve workspace for dump
Screen_info=Gemdos(72,L:34) !Reserve space for palette data
Palette=Screen_info+2
If Routine=0 Or Workspace=0 Or Screen_info=0
Alert 3,"Insufficient memory to|allocate blocks.",1," Abort ",Dummy
Edit !Leave program
Endif
'
Bload "a:*.PRT",Routine !Load printer driver
Return
'
Procedure Memory_free
Status=Gemdos(73,L:Routine)
Status=Gemdos(73,L:Workspace)
Status=Gemdos(73,L:Screen_info)
Return
'
Procedure Screen_dump
@Palette_store
Hidem
Result=C:Routine(W:0,W:Xbios(4),L:Xbios(2),L:Palette,L:Workspace)
                                                    !Initialise routine
Result=C:Routine(W:1,W:Xbios(4),L:Xbios(2),L:Palette,L:Workspace)
                                                    !Execute routine
Showm
Return
'
Procedure Palette_store
Dpoke (Screen_info),Xbios(4) !Store screen resolution
'
For Register=0 To 15
Dpoke Palette+2*Register,Xbios(7,Register,-1) !Get and store palette
Next Register
Return
```

Together `memory_initialise`, `memory_free`, `screen_dump` and `palette_store` provide all the necessary routines to produce a screen dump using a standard Degas printer driver. As a general guide, the basic structure of a program using these procedures should look something like that shown below. The crucial element here is that the memory must be returned to the system before the program ends. Note that the `screen_dump` procedure may be called as many times as necessary during the program before the memory is freed, whereas the two memory procedures should only be called once.

```
@Memory_initialise !Allocate memory and load printer driver
"Draw fractal here
'
@Screen_dump !This may be called many times from this area of the program
'
@Memory_free !Return memory to system
"End of program
```

Loading and Saving Pictures

The simplest way to copy an image from the screen to a disk is to save the 32,000 byte screen memory area as a binary block using the BSAVE command as follows:

```
BSAVE filename$,XBIOS(2),32000
```

where `filename$` contains the filename under which the file should be saved. As in the screen dump program the address of the screen memory is found using XBIOS(2). The complementary BLOAD command can be used at a later date to load the picture back into memory. A typical example of this is:

```
BLOAD filename$,XBIOS(2)
```

Manipulating Degas Format Picture Files

Although the one line method is short and easy it does have its disadvantages. For example, the lack of palette and screen resolution information make it impossible to determine which colour and in which resolution the picture should be displayed. Also, the resultant file does not conform to any of the standard formats used on the ST so cannot be imported into any other programs. The most sensible way to overcome both of these limitations is to use a routine which saves and loads pictures in Degas format. Many commercial programs accept Degas format pictures, which contain all the relevant palette and resolution information.

Degas files consist of two main blocks; the header which contains the palette and resolution data, and the actual screen data. Before saving a Degas picture it is necessary to allocate and fill a block of memory containing the header block. This is done in the `memory_initialise` procedure used in the screen dump suite above, but if the screen dump facility is not required this procedure can be reduced to the more efficient one shown below.

```
Procedure Memory_initialise
Screen_info=Gemdos(72,L:34) !Reserve space for palette data
Palette=Screen_info+2
If Screen_info=0
Alert 3,"Insufficient memory to|allocate blocks.",1," Abort ",Dummy
Edit !Leave program
Endif
Return
```

Useful Routines

The following, similarly simplified, procedure should be used to free the allocated memory once it has been used, before the program ends.

```
Procedure Memory_free
Status=Gemdos(73,L:Screen_info)
Return
```

The procedures to load and save pictures in Degas format are shown below. Note that `degas_save` requires the `palette_store` procedure (listed in the screen dump section above) to be present in the program. The procedures should be called in the usual way, with the filename being given as the argument.

```
Procedure Degas_save(Filename$)
@Palette_store
Open "O",#1,Filename$ !Open file (output)
Bput #1,Screen_info,34 !Write resolution and palette
Bput #1,Xbios(2),32000 !Write picture
Close #1 !Close file
Return
'
Procedure Degas_load(Filename$)
Open "I",#1,Filename$ !Open file (read only)
Bget #1,Screen_info,34 !Read resolution and palette
Bget #1,Xbios(2),32000 !Read picture
Close #1 !Close file
'
For Register=0 To 15 !Load and..
Setcolor Register,Dpeek(Palette+2*Register) !..set palette
Next Register
Return
```

No checks are made in either procedure on the screen resolution. This means that it is possible to load a low resolution picture in high resolution mode, although it will appear severely distorted. Also, the filename extension is not vetted for its validity. A check of this sort would be easy to include and details of how to go about this are given in Chapter 5. Some programs will not accept files whose extension is not correct when checked against the value in the header block so it is important to provide the correct extensions, as follows:

.PI1 for low resolution (320x200, 16 colours)

.PI2 for medium resolution (640x200, 4 colours)

.PI3 for high resolution (640x400, 2 colours)

Useful C routines

Before any ST program that uses GEM is executed the operating system must be initialised and prepared for the task. In GFA BASIC this is done automatically, but in C it is left up to the programmer to provide the relevant initialisation code. This can be very time consuming but in most cases a single piece of set-up code (known as a common shell) can be used for a range of different programs. A compact ST shell for initialising a GEM application in this way is shown in Listing A.1. This loads all necessary header files and calls all the necessary TOS routines before finally calling a function called `program()`. The `program()` function is not included in the shell as it contains the program specific commands. The shell should be entered and saved to disk under a filename such as "C_SHELL.C". It can then be included at the beginning of any program that needs it, with a command like:

```
#include "C_SHELL.C"
```

All of the C programs in this book were written using Megamax Laser C but the portability of C should ensure that they work with any ST compiler that uses standard GEM bindings.

The following list shows the main operations of the shell, and is intended for those with experience of the ST operating system:

❏ Include the necessary header files

❏ Initialise the VDI data arrays

❏ Initialise the AES

❏ Initialise the program as a GEM application

❏ Open a virtual workstation filling the screen

❏ Exit cleanly when the `program()` function returns.

The shell is fairly self-contained but the workstation handle variable, `handle`, might need to be used in the `program()` function as it is essential when calling VDI graphics routines. When dealing with operating system events the application identity, `appl_id`, will also need to be used.

An example program making use of the shell, which simply displays an alert box, is shown in Listing A.2 and a more useful one, to draw the Martin set, is shown in Appendix C. Listing A.2 demonstrates how the shell is included and how the `program()` function performs the main actions of the program.

```
#include <osbind.h>
#include <gemdefs.h>
#include <obdefs.h>
#define DESK 0
extern int gl_apid;
int ap_id;
int contrl[12],
intin[128],
ptsin[128],
intout[128],
ptsout[128];
int phys_handle,
handle;
int work_in[12],
work_out[57];
main()
{
int error;
error = init_all();
if(!error) {
program();
v_clsvwk(handle);
appl_exit();
}
}
init_all()
{
int x;
appl_init();
ap_id=gl_apid;
if (ap_id==-1) return (1);
work_in[10]=2;
work_in[0]=Getrez()+2;
for (x=1;x<10;work_in[x++]=1);
v_opnvwk(work_in,&handle,work_out);
if (handle==0) return (1);
v_clrwk(handle);
return(0);
}
```

Listing A.1: GEM initialisation code


```
#include "C_SHELL.C" /* Tell compiler to include the shell */
program() /*Start of program specific code */
{
form_alert(0,"[3][Example of using GEM|calls from C.][OK]");
}
```

Listing A.2: Program to display an Alert box

You may also find the following mouse pointer manipulation macros useful in fractal programs:

```
#define MOUSE_WAIT evnt_button(1,1,1,&dummy,&dummy,&dummy,&dummy)
#define MOUSE_OFF graf_mouse(256,&dummy)
#define MOUSE_ON graf_mouse(257,&dummy)
```

The above definitions should be included at the head of any program which requires them. They can then be called whenever necessary simply by typing their upper case macro names. For example:

```
MOUSE_WAIT;
```

would cause the program to wait until a mouse button was pressed, while:

```
MOUSE_ON;
```

and

```
MOUSE_OFF;
```

could be used to turn the mouse pointer on and off.

The techniques for manipulating picture files and producing screen dumps in C are so similar to those described above for GFA BASIC, that it is not necessary to repeat them here.

Appendix B

Mathematics in GFA BASIC

This book has been written to appeal to a wide audience of ST users, from those who are accomplished chaos mathematicians, but are unable to incorporate the theory into ST programs, to programmers who are not familiar with the relevant mathematics. This appendix is for reference by those in the second category, and explains the elementary concepts referred to in some of the earlier chapters. If further assistance is required in this area then a good GCSE or O Level mathematics text is recommended, as I have attempted to keep the majority of this book to that level.

Topics are discussed roughly in order of importance, with particular attention being paid to their GFA BASIC implementations.

Simple Algebraic Principles

Algebra encompasses a wide range of fundamental mathematics principles, but probably its most well know aspect is the use of letters to represent numbers. This allows written statements to be expressed in mathematical form. The letters are known as variables, and have similar properties to the variables found in programming languages, except that they may only be one character long. Algebraic variables may be used in all the same ways as normal numbers. For example the algebraic expression shown below involves addition and multiplication.

$$y + y + y = 3 * y$$

The 3*y section would normally be written simply as $3y$, but the GFA BASIC notation adhered to in this book means that the longer equivalent must be used. The equation states that any number (here represented by y) added to itself three times equals 3 multiplied by the number. To use this equation values must be substituted in place of y, as an example the equation when y is four is shown below. Note that in BASIC a number can be assigned to a variable using a command of the form y=4.

$$4 + 4 + 4 = 3 * 4$$

This equation holds for any number, and since it is an equation both sides always equate to the same value. Although this example is of little practical use there are many useful equations expressed in a similar form, such as Pythagorus' theorem.

Variables whose values are always the same are known as constants. By convention different variables are used for different number types. For example x, y, z and i, j, k are usually used to represent variables whereas a, b, c, k and Greek letters such as Π are employed to represent constants. When used alone, z is traditionally used to represent complex numbers, whose constituent parts are usually represented by a for the real part and b for the imaginary one. Apart from having different names there is no distinction between constants and variables in GFA BASIC but in C it is possible to declare variables as being static, meaning that they are treated as constants, which are processed faster.

$$z = a + bi$$

Operator Priorities

In equations with more than two terms it is often important to ensure that the priority of the operators is taken into account when working out the answer. If this is not done then unexpected answers may be produced when carrying out calculations on the computer which, like a calculator, always uses the correct priorities.

For example, the two calculations below yield the same answer, despite the fact that the numbers are in a different order.

```
1 + 2 * 3 = 1 + 6 = 7      NOT    3 * 3 = 9
2 * 3 + 1 = 6 + 1 = 7      NOT    2 * 4 = 8
```

The two calculations equate to the same value because multiplication (and division) has a higher priority than addition (and subtraction), so all multiplying is done first. If the addition needs to be done first brackets must be placed around the addition section in order to increase its priority. The result of the addition of brackets to the last example is:

$$(1 + 2) * 3 = 3 * 3 = 9$$

$$2 * (3 + 1) = 2 * 4 = 8$$

Ratios

A ratio is a way of expressing the relationship between two numbers (or two sets of numbers) without quoting their absolute values. For instance the scale of a model aircraft might be said to be '1 to 48', often written as 1:48. This scale is the ratio of the size of the model to the size of real aircraft, and in this case it states that for a single unit length on the model the equivalent real length is 48 unit lengths. Because the ratio applies to the whole plane it is possible to work out the dimensions of any part of the real aircraft using measurements from the model. For instance if the wing-span of the model is 20cm the wing-span of the real aircraft would be 48*20cm = 960cm or 9.6m.

The colon character (:) is not generally used in maths so ratios are expressed as fractions instead, for example 1:48 would be written as 1/48. This makes manipulation of ratios easier because to scale a length up we just divide the model's length by the ratio, and when scaling down the lengths are multiplied by the ratio. For example the wing-span of the full size aircraft could be calculated as follows (recalling that the model's wing-span is 20cm):

20cm / (1/48) = 960cm

If the fuselage length of the real plane was known to be 816cm the following calculation would be used to calculate the length of the model:

816cm * (1/48) = 17cm

Of course 1/48 could instead be expressed as the decimal number 0.02083. Fractional ratio notation is of particular use when a ratio doesn't contain a 1. For example if the scale of the aircraft above was 3:48 the model's wing-span could not simply be multiplied by 48 as it was for the 1:48 scale.

Fractions as Percentages

Per cent literally means per 100, so percentages are usually given as integers between 0 and 100, for example 93%. However, in maths, and in programming, the per cent character (%) is meaningless in this sense so fractions derived from a ratio are used instead:

Percentage   Ratio   Decimal
93% =   93/100 =   0.93

Using fractions makes percentage calculations much easier, for example to find 93% of 500 we can simply multiply the decimal equivalent of the percentage by 500:

0.93 * 500 = 465

Fractions can be used to represent any percentage and the list below demonstrates most of the common forms.

2%   =   0.02
25%   =   0.25

50%	=	0.50 (usually written as 0.5, without redundant zero)
100%	=	1 (decimal part is totally redundant here)
137%	=	1.37
12.5%	=	0.125 (decimal part may be extended indefinitely)

Graphs and Co-ordinate Geometry

The only way of creating pictures of any kind in mathematics is by using graphs, and fractals are no exception.

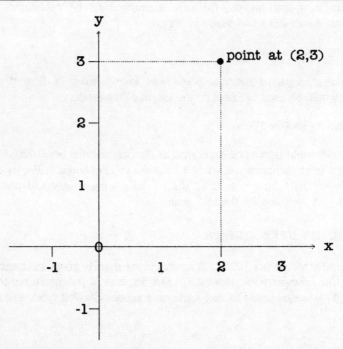

Figure B.1: Typical two dimensional graph

A typical graph is a two-dimensional plane on which points and lines can be plotted, like the one shown in Figure B.1. Each point on the plane has a unique two part co-ordinate associated with it, consisting of the horizontal and vertical positions of the point. These positions are given numerically with reference to a pair of graticuled lines, called axes, which run through the graph at right angles to one another. The horizontal axis is generally called the x axis and gives the x position of points on the graph. Similarly the y position of points is read from the vertical y axis. Thus the example point in Figure B.1 can be said to be at the position where $x=2$, $y=3$. For neatness the x and y values are normally put together in a single co-ordinate packet of the form (x,y). In the current example, this would be written as $(2,3)$.

Mathematics in GFA Basic 183

A typical application of a two dimensional graph is the display of time series charts, where a quantity (on the vertical axis) is plotted against time (on the horizontal axis). A selection of time series graphs can be found in Chapters 2 and 3.

The graph used in the examples above was a standard two dimensional one, but one dimensional graphs with a single axis and three dimensional ones with three axes (x, y and z) are also used in this book. These all work on exactly the same principle. In fact as many dimensions as necessary can be used as long as there is one axis to represent each dimension.

Note that x and y are the normal variable names for the horizontal and vertical positions of a point but any variables can be used. To find out which variable represents which direction you can simply look at the axes, as these are labelled with the variable names.

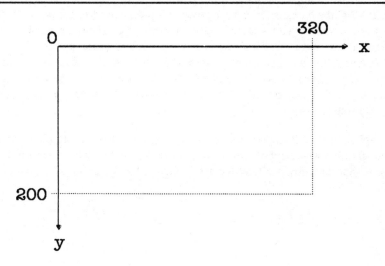

Figure B.2: The ST's screen as a graph

The ST's screen can be thought of as a graph, with the axes shown in Figure B.2. Graphs can be drawn on the screen by specifying point and line positions using commands like PLOT and DRAW. Unfortunately, however, the range of the screen axes is fixed, so graphs need to scaled up and down in order to make maximum use of the screen area. Also, the y axis is inverted making direct graph plotting rather difficult. A bonus of the ST's screen (in low and medium resolution modes) is that in addition to a position co-ordinate each point also has a colour associated with it. This allows more information to be displayed in a smaller space, for example it can be used to represent the third dimension of the Lorenz attractor.

Indices

Commonly known as *powers*, indices are a very useful mathematical tool, used in many applications. An example of index notation is shown below. The number shown in normal sized type is called the base and the smaller superscripted number is known as the index.

2^4

The equivalent GFA BASIC notation (used in this book), where the base and index are separated by a ^ character, is:

2^4

We say that this example shows 2 to the power of 4, or that 2 is raised to the power of 4. In BASIC the ^ character can effectively be pronounced *to the power of*. Certain indices have particular names associated with them. For example a number raised to the power of 2 is said to be squared and a number raised to the power of 3 is said to be cubed, e.g. 5^3 is pronounced *five cubed*. It is also common to say the a number is *the square* of another number. For example 16 is the square of 4, because 4^2 equals 16. Similarly 125 might be called the cube of 5 (details of how to calculate these values are given below).

In its simplest form index notation is basically a quick way of expressing calculations which would otherwise be quite tiresome to write. The index denotes how many times the base is multiplied by itself, so the actual value of 2^4 can be calculated as shown below.

2^4   =   2 * 2 * 2 * 2   =   16

Similarly:

10^2   =   10 * 10   =   100
x^4   = $x * x * x * x$

Square Roots

The opposite action to raising a number to a power is called rooting a number. The only type of rooting used in this book is square rooting, which has the opposite effect to raising a number to the power of 2. An example of this is shown below. Note that the BASIC SQR syntax is used to denote square rooting in this book.

SQR(64)   =   8   (because 8^2 = 8*8 = 64)

The MOD Operator

The conventional use of the BASIC MOD operator is for finding the remainder after integer division has been performed. However, it is also very useful for keeping numbers within certain ranges without giving adjacent numbers the same value. This technique is used in Mandelbrot and Julia programs when converting iteration numbers of unknown range to the 16 colour range of the ST's low resolution screen mode.

An everyday use of MOD can be found in the display on a 24 hour digital watch. The part of the display showing hours is incremented every hour, on the hour, but when the end of the 23rd hour occurs (at 23:59:59) the hour value is set back to zero, from where the process begins again. The hour value is said to be modulus 24 because it has 24 possible values (0 to 23 inclusive). This modulus solution must be used here in order to prevent parts of the time from becoming too large to handle. The only alternative is to stop the hour value when it gets to 23, which would be equally unworkable.

The 24 hour clock face in Figure B.3 graphically demonstrates a practical example of MOD at work. The hour hand of the clock can rotate an infinite number of times without ever stopping or leaving the range of 0 to 23, but no two adjacent hours ever have the same value.

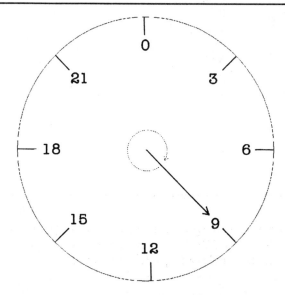

Figure B.3: 24-hour clock face showing the action of MOD 24

The clock face analogy can be used to demonstrate all instances of modulus arithmetic. For example Figure B.4 shows the application of MOD 4. This clock has four possible values, 0 to 3 inclusive. Note that the range always begins at 0, an offset would need to be added to the result if a different base was required. Table B.1 demonstrates the use of the four hour clock by showing the value of a number, x, against (x MOD 4).

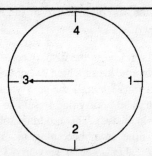

Figure B.4: *Four-hour clock face demonstrating MOD 4*

Table B.1: *The action of MOD 4*

x	(x MOD 4)
0	0
1	1
2	2
3	3
4	0
5	1
6	2
7	3
8	0
9	1
etc.	etc.

The following GFA BASIC program can be used to demonstrate the MOD operator. On each cycle of the DO...LOOP loop a variable, x, is incremented and then printed on the screen next to the corresponding value of (x MOD 4).

```
X=0
Do
Print X,(X Mod 4)
Inc X
Loop
```

Angles – Degrees and Radians

An angle describes the size of a turn. Angles are usually expressed in degrees, where a full circle consists of 360 degrees, a semi-circle has 180 degrees and so on. However, in GFA BASIC the more mathematical radian equivalent is used, where angles are normally expressed in terms of the constant, PI. PI is a common mathematical constant known to GFA BASIC, and is equal to approximately 3.1416.

Degrees	360°	180°	90°	45°	30°
Circle Portion	●	◐	◢	◣	◣
Angle					
Radians	2*PI	PI	PI/2	PI/4	PI/6

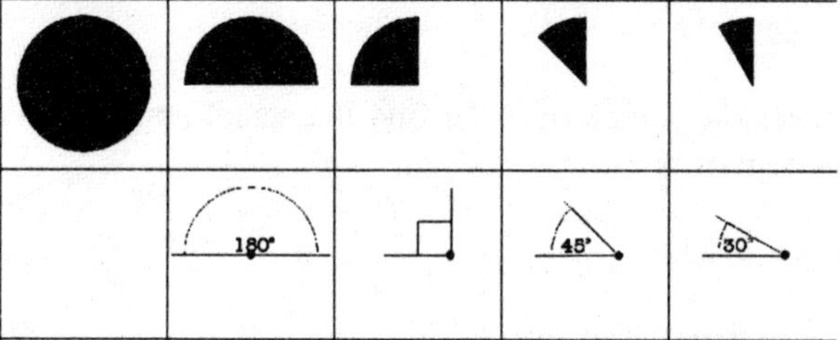

Figure B.5: Common degree to angle conversions

Figure B.5 shows the equivalent degree and radian values for some common angles. In the case of more general values the two equations below can be used to convert between radians and degrees.

degree_value   =   180 * (radian_value/PI)
radian_value   =   degree_value * (PI/180)

Peculiarities of Right-angled Triangles

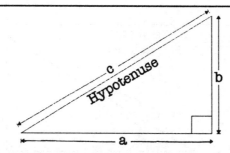

Figure B.6: A typical right-angled triangle

A right angled triangle is one which contains a right-angled (90 degree) corner. An example of a triangle of this type, with the right-angle and hypotenuse marked, is shown in Figure B.6. The hypotenuse is just the mathematical name for the longest side, which can always be found directly opposite the right-angle.

Pythagoras' Theorem

This simple law states that the square of the longest side of a right-angled triangle (the hypotenuse) is equal to the sum of the squares of the two other sides. Taking the triangle in Figure B.6 as an example we can form the following equation:

$c^2 = a^2 + b^2$

This only gives the square of c. To find the actual length both sides of the equation must be square rooted:

$c = SQR(a^2 + b^2)$

Simple Trigonometry

There are three useful equations which allow the length of a right-angled triangle's sides to be calculated from one of its angles and vice-versa. These are shown in Figures B.7(a) to B.7(c).

Figure B.7: Application of the three common trigonometrical ratios

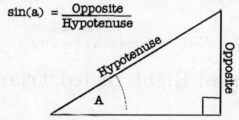

(a)

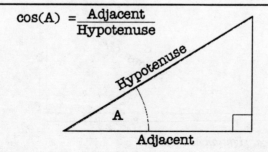

(b)

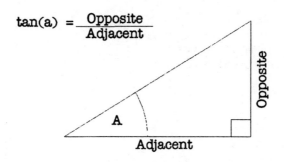

(c)

Note that *opposite* is used to refer to the length of the side opposite the angle, A, and *adjacent* is the length of the non-hypotenuse side which touches the angle. Sin, Cos and Tan are mathematical functions, which are built into BASIC. The relevant syntax for these functions is:

result = SIN(angle)

result = COS(angle)

result = TAN(angle)

Note that in GFA BASIC the angles must be given in radians.

Appendix C

Using Other ST Languages

If GFA BASIC is not your preferred programming language you may want to know how to convert the listings in this book. Because GFA BASIC specific commands have been avoided where possible and because the majority of the programs consist mainly of standard maths and loops, conversion to other BASICs is relatively easy.

It would be impossible to discuss all the differences between the popular ST languages in this appendix, so instead only the most important differences for fractal work are mentioned. To find these differences I wrote a typical fractal generation program under each one. This program performed the same action in each language, it carried out the first 25,000 iterations of the Martin equation (see Chapter 8) with constants $a=45$, $b=2$ and $c=-300$, and plotted the resulting fractal. This program was chosen because it is very typical of fractal programs, involving loops, maths and the plotting of points. Each version of this Martin program was timed before and after being compiled (if relevant) to give an indication of the suitability of each language for fractal work. The resulting times are shown in order in Table C.1. Note that the C language is not included in the table as speed varies enormously between compilers.

Table C.1: *Performance comparison of popular ST BASICs. All programs were as similar as possible and all variables were defined as floating point numbers.*

Language	Time for 25 000 iterations of Martin equation (seconds)
Compiled GFA BASIC 2.0	55
Compiled Hisoft Basic 1.10	59
Interpreted GFA BASIC 2.0	90
Compiled STOS 2.4	98
Interpreted STOS 2.4	155
ST BASIC	760

Hisoft BASIC

Hisoft BASIC is intended to be used for the same type of applications as GFA BASIC and therefore contains many of the same commands, although a few inevitably have different syntax. Notable differences are the use of PSET instead of PLOT, MOUSE -1 instead of HIDEM and DO...LOOP UNTIL instead of REPEAT...UNTIL. Hisoft BASIC is more structured than GFA BASIC, and allows constants and other special variables to be defined, which is very useful in chaos programming. When converting the programs in this book to Hisoft format you should use such special definitions wherever possible to improve speed and clarity.

STOS BASIC

The key point to remember when converting programs to STOS is that all non-integer variables must be suffixed by a hash character (#). Because STOS is geared to game creation, where floating point numbers are rare, it treats all variables as integers unless instructed otherwise. In most other respects STOS has the same syntax as GFA BASIC except for a few notable exceptions, the most prominent of which is the use of line numbers rather than named procedures. The many extra non-standard commands provided by STOS make some tasks much simpler. For example a screen picture can be saved in Degas format simply by using a one line SAVE command. STOS does not support GEM, but provides similar facilities of its own.

ST BASIC

Considering that the ST BASIC version of the 25,000 iteration Martin program takes over eight times longer than the equivalent GFA BASIC one (see Table C.1) it is unlikely that anyone would seriously consider using ST BASIC for fractal work. A discussion of the differences between the two BASICs is therefore inappropriate.

The C Language

Although specific applications for C have been cited throughout this book, no example code has been given. Originally I had intended to present all the example programs in both C and GFA BASIC, but this would only make the principles of chaos theory harder to grasp. Also, it was assumed that most C programmers would be able to read BASIC and should therefore be able to convert the programs with relative ease. A C version of the Martin program from Chapter 8 is provided here to assist C programmers who are unfamiliar with GFA BASIC and as an example to GFA BASIC programmers who would like to progress to writing programs in C.

The program is shown in listing C.1. It was written using Megamax Laser C but should be accepted by any standard compiler. Note that the C shell given at the end of Appendix A, and the "MATH.H" header file should be available to your compiler when compiling this program. The mon variable should be set to 1 when running in low resolution and 2 when using high resolution.

```c
#include "C_SHELL.C" /* Include the shell */
#include <MATH.H> /* May need #include "math.h" */
/* on some compilers */
/* The following macros have been included from appendix A */
#define MOUSE_OFF graf_mouse(256,&dummy)
#define MOUSE_ON graf_mouse(257,&dummy)
program() /* Function containing program-specific code */
{
/* Declare floating point variables */
float x=0,
y=0,
xnew,
ynew;
/* Declare integer variables */
int i=0, /* iteration number */
dummy, /* dummy variable */
point[2], /* pixel co-ordinate */
button=0, /* mouse button state */
sign; /* sign of number */
/* Declare constants for required set */
static int a=45,
b=2,
c=-300;
/* Declare integer monitor constant */
static int mon=2; /* mon=1 for low-res */
/* Turn mouse pointer off using macro */
MOUSE_OFF;
/* Clear screen ('handle' is provided by the shell) */
v_clrwk(handle);
/* Enter main loop (terminated by mouse button press) */
while (button==0) {
/* Check mouse button state */
vq_mouse(handle,&button,&dummy,&dummy);
sign=0;
if (x>0) sign=1;
if (x<0) sign=-1;
xnew=y-sign*sqrt(dabs(b*x-c));
ynew=a-x;
x=xnew;
y=ynew;
/* Plot single point */
/* Low level line A calls would be quicker, */
/* but the syntax varies between compilers */
point[0]=(150+x*0.4)*mon;
point[1]=(100-y*0.4)*mon;
v_pmarker(handle,1,point);
/* Increment i */
i++;
}
/* Use macro to turn pointer back on */
MOUSE_ON;
}
```

Listing C.1: A C version of the Martin program

Appendix D

Use of ST Peripherals

There is now a wide variety of expansion options and peripherals available for the Atari ST range of computers, but to enable all ST owners to use this book the major programming examples have been written around the basic 520ST. This system has 512k of RAM and an internal floppy drive with a minimum formatted capacity of about 400k. However, fractal generation will benefit from some peripherals, this chapter introduces the techniques necessary to adapt existing programs to make the most of such peripherals.

Monitors

All the listings contained in this book work on both low resolution colour and high resolution monochrome monitors, but although the programs will work in low resolution on a standard television set the image will not be as sharp as those displayed on a dedicated ST monitor.

When preparing this book I used the Atari SM124 high resolution monochrome monitor, because it is the only ST monitor which can cope with the high resolution graphics mode. This mode's resolution of 640x400 pixels makes it ideal for displaying almost all types of fractal and screen dumps from it are quite acceptable for publishing.

There are a number of colour ST monitors on the market, all allowing the use of both low (320x200, 16 colours) and medium (640x200, 4 colours) resolution screen modes. Medium resolution on such monitors is of a similar standard to that of the high resolution display, except that each pixel is twice the height. Medium resolution is not used is this book because of its poor appearance on standard televisions and its unique aspect ratio. However, the definitions for the `fdraw` and `fplot` procedures shown below can be used to force many of the listings to run in medium resolution mode if this is desirable.

```
Procedure Fplot(X,Y)
Xp=X*Monitor
Plot Int(Xp)+Int(Frac(Xp)*2),Int(Y)+Int(Frac(Y)*2)
Return
'
Procedure Fdraw(X,Y)
Xp=X*Monitor
Draw To Int(Xp)+Int(Frac(Xp)*2),Int(Y)+Int(Frac(Y)*2)
Return
```

Low resolution is rather poor for displaying monochrome fractals, but its range of 16 colours is an obvious advantage when plotting the Mandelbrot and Julia sets. Also, its large pixels look steady and well defined even on a standard television set.

Printers

If the fractals you create are intended to reach a wider audience than fellow ST owners, a printer will become a necessity. Figures D.1(a) to D.1(c) show the difference between the output of the three main monochrome printing methods.

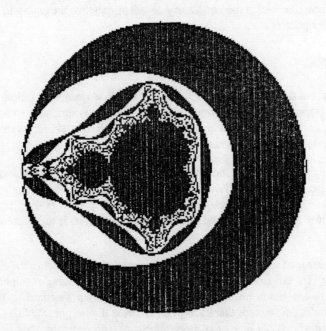

Figure D.1a: Low-resolution screen dump to 9 or 24 pin printer (320x200)

Use of ST Peripherals 195

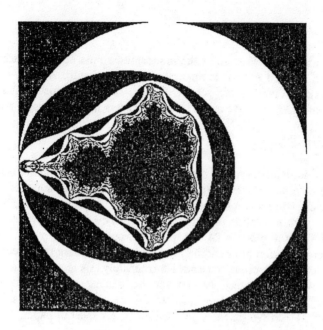

Figure D.1b: High-resolution screen dump to 9 or 24 pin printer (640x400)

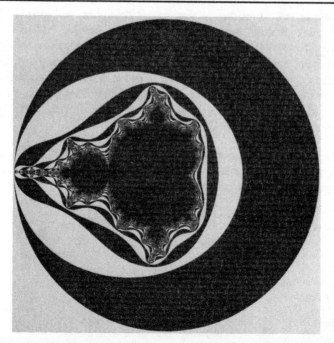

Figure D.1c: Very high resolution image produced on a 24 pin printer by Listing D.1 (1416x1416 pixels)

Nine Pin

The most common ST printers are of the monochrome nine pin dot-matrix type, and with the help of a good screen dump routine (see Appendix A) they can quickly produce an accurate copy of the screen display, with the same resolution as the screen. Colour may also be represented on such printers using shades of grey.

Twenty-four Pin

Twenty-four pin dot-matrix printers are now becoming far more numerous. They can produce output of around 180 dpi (dots per inch) which is far superior to nine pin printers. However, when dumping screen images the resolution of the dump is of course limited by the resolution of the screen, and since nine pin printers can easily manage 640x400 pixel plots a 24 pin printer would appear totally unnecessary. However, higher resolution plots can be produced by by-passing the screen altogether and writing directly to the printer instead. Unfortunately this technique can only really be used to draw fractals which are created by placing pixels in rows, like the Feigenbaum diagram and Mandelbrot/Julia sets. When plotting such fractals near laser quality output of 1416x1896 pixels is possible on an A4 sheet of paper.

The program given in Listing D.1 demonstrates the 24 pin printing method for the Mandelbrot set. After requesting information on the section of the set that you want to plot, the program will print that section with a resolution of 1416x1416 pixels. It sounds like a quick process, but remember that there are 1414x1416 = 2,005,056 pixels (compared to 64,000 or 256,000 when plotted to the screen) and each pixel must be tested using one of the time consuming methods discussed in Chapter four. Also, the iteration ceiling must be fairly high in order to differentiate between pixels which belong to the actual Mandelbrot set and those that do not. Many calculations are necessary and plots taking over 24 hours to produce are not uncommon when running the program in interpreted GFA BASIC.

This program has been written specifically for the Epson LQ-500, but it should work with any compatible 24 pin printer. However, it must be stressed that I have not tested the program on any other printers.

```
' 1416 x 1416 Mandelbrot Direct To Printer (V1.22SX)
' (c)1991 Conrad Bessant
'
Gosub Get_values
'
Lprint Chr$(27);Chr$(51);Chr$(24);  !Set line feed size
For Yl=0 To 1415 Step 24
Lprint
Lprint Chr$(27);Chr$(42);Chr$(39);Chr$(136);Chr$(5);  !Tell
printer that 1416 columns of 180 dpi graphics are coming
For X=0 To 1415
For Byte=1 To 3
```

```
Value=0
For Bit=0 To 7
Y=1416-(Y1+Byte*8-Bit-1)
I=X/N+Imin
J=Y/N+Jmin
A=0
B=0
T=0
While A*A+B*B<4 And T<120
T=T+1
C=A
A=I+(C*C-B*B)
B=J+2*C*B
Wend
If T Mod 2=0
Value=Value+(2^Bit)
Endif
Next Bit
Lprint Chr$(Value);
Next Byte
Next X
Next Y1
Lprint Chr$(10)
Lprint Chr$(12)
'
Procedure Get_values
Print "Please enter required parameters:"
Print
Input "Minimum value of a (amin):",Imin
Input "Maximum value of a (amax):",Imax
Input "Minimum value of b (bmin):",Jmin
Print "Please wait..."
S=Imax-Imin
N=1416/S
Return
```

Listing D.1: Program to output the Mandelbrot set to an Epson LQ-500 24 pin printer

It would be tedious to explain this program in detail, but the key difference between this and the similar screen oriented ones given in Chapter 4 is the order in which the pixels are plotted. Because the screen is effectively a random access device it is possible to plot pixels in the most convenient way, usually in vertical columns drawn from left to right. Conversely, the printer expects data to be sent in a precise format and as a result the order in which the pixels are plotted has to be tailored to suit the printer. In addition, the printer is fairly unintelligent so we need to send the pixel information in low level bits, with control codes governing the way the data is interpreted. The way in which the image in built up on the LQ-500 is shown in Figure D.2.

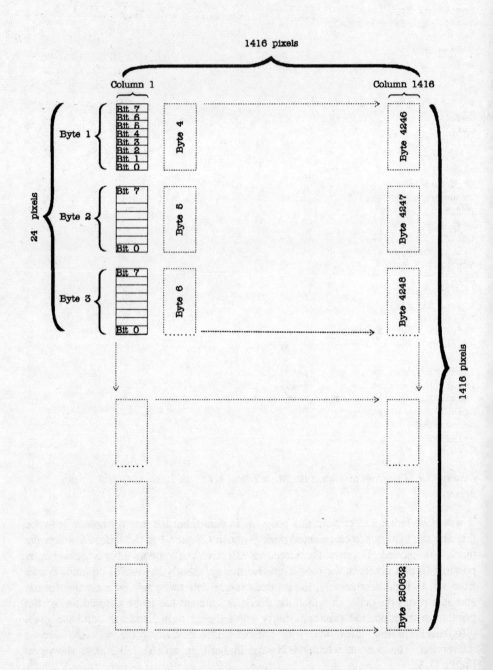

Figure D.2: Structure of a 180dpi picture on the Epson LQ-500

Colour

Obviously a colour printer would be desirable for producing some fractals, and the resolution of such printers can normally cope with ST screen dumps up to 640x400 pixels in size. However, colour printer drivers for the ST are scarce and you may be forced to write your own. If you have a colour printer and a compatible Degas printer driver you can incorporate it in your GFA BASIC chaos programs using the methods discussed in appendix A.

Plotters

Due to the constraints imposed by their design, plotters can realistically be used only for drawing fractals consisting mainly of lines. The fact that plotters physically draw with real pens means that they can produce very high resolution Lorenz attractors, fractal plants and landscapes if accessed directly (without plotting to the screen). Other fractals, such as the Mandelbrot set, are very difficult to produce on a plotter because the drawing of single dots is slow and screen dump programs for plotters are rare.

Extra RAM

Fractal generation is a relatively memory efficient process because the programs are short and most of the information is stored in the 32k screen area reserved by GEM at all times. Extra memory is only really useful for animating fractals. On a two megabyte machine, for example, it would be possible to create 36 three dimensional pictures (frames) of the Lorenz attractor, each one rotated 10 degrees in relation to the previous one. After being drawn on the screen (not necessarily with any great speed) each frame could be stored in RAM. Once all the frames had been stored they could then be copied onto the screen in quick succession to create the illusion of a rotating three dimensional Lorenz attractor. Having tried this technique on the Acorn Archimedes I can testify that the end result is very effective and dispels any doubt about the Lorenz line intersects itself.

From Chapter 6 you may recall that extra memory is also useful when drawing very detailed fractal plants.

Bibliography

James Gleick, *Chaos: Making a New Science* (Viking Press, 1987)

This is the most popular introduction to chaos, because of its agreeable style and accessibility. It describes a wide variety of fractals and chaotic processes, backed up by a host of real world chaos examples. The addition of historical information and interviews with people such as Mitchell Feigenbaum help create a useful insight into the modern scientific community, which is very enlightening if you have even considered becoming part of it. Unfortunately, few mathematical details are given, making it difficult to recreate the experiments discussed.

Karl-Heinz Becker and Michael Dorfler, *Dynamical Systems and Fractals: Computer experiments in Pascal* (Cambridge, 1989)

In stark contrast to *Chaos* this book is composed mainly of technical information such as equations and programs, with little historical background or real world examples to relate to. The programs are written in Pascal for the Apple Macintosh, but are explained well enough to make conversion to other computers relatively easy (as long as you can read Pascal).

It contains several unexpected topics, such as a chapter entitled *Journey to the Land of Infinite Structures* in which the authors take the reader on a flying trip around the Mandelbrot set. However, the book does not suffer from these interludes or from the fact that it has been translated from German and the resulting mix of hard information and some lighter moments make this the ideal companion for amateur scientists who wish the carry out their own chaos experiments.

Heinz-Otto Peitgen and Peter H. Richter, *The Beauty of Fractals*, (Springer, 1985)

The wealth of fractal pictures (many in colour) adorning *The Beauty of Fractals* has apparently made it the first ever coffee table mathematics book. However, there is some detailed theory buried among the pictures and it has to be said that such high quality fractals would be hard to find anywhere else (although some of these are reproduced in *Chaos*). This is really the sort of book to get out of the library, as it doesn't contain much in the way of reference material and is quite expensive. However, if your coffee table is looking slightly bare then it may be worth getting your own copy.

Benoit Mandelbrot, *The Fractal Geometry of Nature* (Freeman, 1982)

Mandelbrot washes down the sometimes difficult, and now famous, theory with his personal recollections, comprising details of the equipment he was using and what he was trying to achieve when he formulated certain theories. Some of his original output is also included, most notably the first ever printout of the Mandelbrot set. Being one of the first books on chaos, people are inclined to dismiss this as being nothing more than a historic artifact. However, much of the work is still very relevant. In fact a great deal of the modern chaos literature is based on theories derived from this and Mandelbrot's other book, *Fractals: Form, Chance, and Dimension* (Freeman, 1977).

Predrag Cvitanovic, *Universality in Chaos* (Adam Hilger, 1989)

This is a collection of notable scientific papers on chaos published over the last 20 years. Such papers provide the information and inspiration for almost all chaos books, and are where the now famous discoveries were first published. Pieces by various authors are featured including Edward Lorenz, Mitchell Feigenbaum and Robert May. Although the papers are only fully comprehensible to science graduates useful information for chaos programming on the ST can be gleaned.

K. Sullivan, *The Big Red Book of C* (Sigma Press, 1983)

In order to utilise the full power of the ST, programs should really be developed using a C compiler, the structure and speed of the language make it ideal for chaos work and if *Chaos, Fractals and the Atari ST* has urged you to make the step up to C you really need a suitable tutorial text from which to learn it. Although I know of no such books written specifically for the ST, there are many more general works, of which The Big Red Book represents a good value example. All the main aspects of the language are described but supplementary information, specific to the ST, would be required before embarking on any major programming project.

If you wish to continue chaos programming using GFA BASIC a number of books are available to help you master the general programming techniques. At least two such works are published by GFA themselves and many third-party products are available (ST magazines usually carry adverts and reviews of new titles).

It is important to note that this is only a very small selection of relevant literature and that much more material is available in the form of scientific papers, articles and other books. Such material is often highly mathematical, so I would not advise moving up to these books right away. However, most of the chaos books above have their own unique bibliographies which will give more specialised titles, hopefully leading you gently in the direction in which you want to go.

INDEX

A
ABS, 145
Alert box, 79
Angular measure, 187
Art and chaos, 158
Artificial intelligence, 162
Aspect ratio, 15
Attractors, 21
    pi, *see* pi attractor
    Lorentz, *see* Lorentz Attractor
    strange, *see* Strange Attractor
Axes, graphical, 182

B
Barnsley, Michael, 157
Bifurcation, 26
Biology, 15
Brahe, 1
Bursting phenomenon, 150
Butterfly effect, 35, 39, 56, 151, 158

C
C language, 4, 122, 191
C language routines, 176
C-Curve, 114, 115
Cell culture, 145
Chaos, future of, 156
Chaos, general, 2
Chaotic regions, 23, 26
`check_res`, 12, 170
Chemistry, 42

Clarke, Arthur C., 157
Clouds, 154
Co-ordinate geometry, 182
Coastlines, 67
Colour, use of, 28, 44, 56, 143, 152, 200
Complex numbers, 59
Complex number, modulus of, 61
Contamination constant, 19
Contours, 74

D
Data compression, 156
Degas format, 80, 174, 175
Degas printer driver, 171
Deterministic processes, 9, 48, 145, 159
Differential equations, 31

E
Ecology, 16
Economics, 151
Equilibrium states, 26
Evolution, 159
Faulting (landscapes), 135

F
`Fdraw`, 171
Feedback, 31, 46, 146
Feedback, mathematical, 20
Feigenbaum diagram, 15, 24, 99, 139, 153, 163
Feigenbaum landscape, 139

Feigenbaum, Mitchell, 202
Ferns, 99
File selector, 83
Fluid dynamics, 29
Fplot, 170
Fractals, 13, 66, 67, 99, 114
    dimension of, 68, 99
    3D, 139
    general features of, 3
    landscapes, 123
    Martin, 146
Function library, 6, 169

G
Galilileo, 1
GEM, 76, 176
GEM menu bar, 77
GFA BASIC, 4, 42, 76, 100, 154, 190
GFA BASIC compiler, 106
Graphs, 182

H
Hidden line removal, 126, 127
Hisoft BASIC, 4, 191

I
Indices, 184
Inheritance (landscape generation), 132
Initial conditions, 39
Initial conditions, sensitive dependence on, 53
Internal structure, 75
Isometric 3D, 44, 123 - 125
Isometric drawing, 123
Iteration ceiling, 52, 139, 140
Iteration, 10, 139, 147
Iterative processes, 10, 106, 145

J
Julia landscapes, 75
Julia set explorer, 75
Julia sets, 72 - 74, 99

K
Kepler, 1
Koch curve, 116
Koch curve, gradient of, 120
Koch flake, 116

L
Landscapes, 133
    dynamics of, 123
    Feigenbaum, 139
    fractal, 123
    Mandelbrot, 139, 140
    pseudo-natural, 131
Logo language, 102
Lorentz attractor, 40, 145, 200
Lorentz, Edward, 29, 202
Lorentz weather model, 29

M
Magnification, of curve, 76, 120
Mandelbrot landscape, 139, 140
Mandelbrot set, 45, 51, 53, 58, 59, 60, 139, 156
    anaemic, 69, 70
    internal structure of, 71
    manipulation of, 63
    two-headed, 167
    spiky, 164
Mandelbrot, Benoit, 45, 202
Martin fractals, 146
Martin, Barry, 145
Mathematical notation, 5
Mathematics, 179
May equation, 16, 19
May, Robert, 16, 202
Menger sponge, 15
Menu handling, 78
Meteorology, 29
MOD function, 56
MOD operator, 185
Modulus, of complex number, 61
Money markets, 39
Monitor compatibility, 5
Monitors, 194
Monte Carlo method, 159

N
Natural selection, 159
Nature, imitating, 98
Newton, 1

O
Operator priorities, 180
Oscillation, 21, 22

P

Pascal language, 201
Peripherals, 194
Photosynthesis, 99
Pi, calculation of, 159
Pi attractor, 162
Pictures, loading and saving, 174
Plants and shrubs, structures of, 98, 101, 102
Plotters, use of, 200
Politics and chaos, 39
Population dynamics, 15
Printers, 195
Procedure library, 6, 13, 169
Pythagoras' theorem, 49, 188

R

RAM, 200
Random processes, 9, 131
Ratios, 180
Ray tracing, 143, 158
Real and imaginary numbers, 59
Recursion, 105
REM, 7
Remarks, 7
Rössler attractor, 42

S

Screen dumps, 7, 171
Self-similarity, 13, 26, 66
Shading, use of, 143
Sierpiñski triangle, 9, 11, 13, 20
Smoke, 154
Snowflake, structure of, 114
Sound, use of, 26, 153
Sparks, simulation of, 154
SQR, 145, 145
Square roots, 184
ST BASIC, 4, 191
STOS BASIC, 4, 42, 191
Strange attractor, 42

T

Time series, 20
Trees, structure of, 110, 143
Trigonometry, 188

V

Variables, integer, 77
Viruses, 16

W

Waitmouse, 171
Weather forecasting, 29, 151

Z

Zooming in, 63, 76